Two Manuals Combined into One Book

FM 7-85 Ranger Unit Operations

And

Soldier's Handbook for Individual Operations Survival in Cold Weather Areas

Department of Defense

Pentagon
Publishing

TRAINING CIRCULAR
No. 21-3

HEADQUARTERS
DEPARTMENT OF THE ARMY
Washington, DC, 17 March 1986

SOLDIER'S HANDBOOK FOR INDIVIDUAL OPERATIONS AND SURVIVAL IN COLD-WEATHER AREAS

TABLE OF CONTENTS

PREFACE.

CHAPTER 1. INTRODUCTION TO COLD-WEATHER OPERATIONS

1-1. Cold Conditions 1-1
1-2. Effects of Cold Weather on Military Operations 1-1
1-3. Overcoming the Cold 1-2
1-4. Positive Leadership and the Right Attitude 1-3

CHAPTER 2. PERSONAL CLOTHING AND EQUIPMENT

2-1. Individual Clothing 2-1
2-2. Cold-Wet Versus Cold-Dry 2-2
2-3. The Cold-Weather Uniform 2-3
2-4. Extended Cold-Weather Clothing System (ECWCS) (TEST) 2-5
2-5. Tips on Wear and Maintenance of Clothing 2-6
2-6. Cold-Weather Equipment 2-6
2-7. Load-Carrying Equipment 2-8
2-8. Over-the-Snow Movement Equipment 2-9
2-9. Miscellaneous Equipment 2-11

CHAPTER 3. TENTS AND HEATING EQUIPMENT

3-1. General 3-1
3-2. Tent Group Equipment 3-1
3-3. AHKIO 3-2
3-4. Packing 3-2
3-5. Ten-Man Arctic Tent 3-2
3-6. Building Arctic Tents 3-3
3-7. Instructions for Pitching Tents 3-4
3-8. Yukon Stove 3-5
3-9. Precautions 3-6
3-10. Squad Stove M1950 3-7

*** This publication supersedes TC 21-3, 30 September 1974.**

CHAPTER 4. SMALL ARMS AND AMMUNITION

4-1. General 4-1
4-2. Care, Cleaning, and Maintenance 4-1

CHAPTER 5. RATIONS AND DIET

5-1. General 5-1
5-2. Rations 5-1
5-3. Liquids 5-1

CHAPTER 6. HYGIENE AND FIRST AID

6-1. General 6-1
6-2. Hygiene Tips for the Soldier 6-1
6-3. Hypothermia 6-2
6-4. Frostbite 6-3
6-5. Dehydration 6-5
6-6. Carbon Monoxide Poisoning 6-6
6-7. Prevention 6-7
6-8. Sunburn 6-8
6-9. Tent Eye 6-8
6-10. Trench Foot/Immersion Foot 6-8
6-11. Constipation 6-9
6-12. Diarrhea 6-9
6-13. Heat Cramps 6-9
6-14. Heat Exhaustion 6-10

CHAPTER 7. BIVOUAC ROUTINE

7-1. General 7-1
7-2. Bivouac Site 7-1
7-3. Site Selection 7-1
7-4. Reconnaissance of Bivouac Site 7-3
7-5. Trail Plan 7-3
7-6. Occupation 7-4
7-7. Bough-Cutting Areas 7-4
7-8. Garbage 7-4
7-9. Latrine 7-4
7-10. Water Points and Snow Areas 7-5
7-11. Storage 7-5

CHAPTER 8. INDIVIDUAL MOVEMENT

8-1. General 8-1
8-2. Skis 8-1
8-3. Snowshoes 8-1

CHAPTER 9. LAND NAVIGATION

9-1. General 9-1
9-2. Skis 9-1
9-4. Distance and Direction 9-1
9-5. Measuring Distances 9-2
9-6. Determining Direction 9-2
9-7. Navigation by Dead Reckoning 9-4
9-8. Hints for the Navigator 9-5

CHAPTER 10. SURVIVAL

10-1. General 10-1
10-2. Action When Lost 10-1
10-3. Improvised Shelters 10-2
10-4. Food 10-8
10-5. Useful Knots 10-12

REFERENCES References-1

GLOSSARY Glossary-1

PREFACE

This handbook is an aid for soldiers in cold climates. However, expertise in dealing with cold-weather operations cannot be gained just from reading this handbook. To develop successful cold-weather field techniques, application training and operational experience are necessary. Additional information is available in FM 31-70, FM 31-71, FM 90-11 (TBP), and FM 9-207.

Military history of operations in northern areas shows that man's success or failure is measured by his regard for the dominant characteristics of a cold-weather environment. For a soldier to become a cold-weather expert, he must learn the techniques and rules needed to fight and live in a cold-weather climate.

The proponent of this publication is HQ TRADOC. Submit changes for improving this publication on DA Form 2028 (Recommended Changes to Publications and Blank Forms) and forward it to Commandant, US Army Infantry School, ATTN: ATSH-B, Fort Benning, Georgia 31905-5593.

Unless otherwise stated, whenever the masculine gender is used, both men and women are included.

CHAPTER 1

Introduction to Cold-Weather Operations

"Soldiers in pairs look after each other; be faithful to your mate; be loyal to your leaders."

Anonymous

1-1. COLD CONDITIONS

COLD-WET

Cold-wet conditions occur when temperatures are near freezing and high and low temperatures cause alternate freezing and thawing. These conditions can be accompanied by wet snow and rain, causing the ground to become slushy and muddy. Soldiers need clothing that has a waterproof or water-repellent, wind-resistant outer layer, and an inner layer with insulation that provide protection in moderately cold weather (above 14°F). **Waterproof footwear is essential.**

COLD-DRY

Cold-dry conditions occur when average temperatures are lower than 14°F. The ground is usually frozen and the snow dry. Low temperatures plus wind increase the need for protection of the entire body. For these conditions, soldiers need clothing that will provide insulation for a wind-chill factor of -80°F. For temperatures below 14°F, white vapor barrier (VB) boots effective down to temperatures of -50°F must be worn.

WIND CHILL

Wind chill is a measure of the combined effects of wind and temperature. Temperature alone does not give a true indication of the impact of the cold. In order to effectively gauge the difference between temperature and the impact of the cold, a wind-chill scale must be used. The chart in Appendix A is a simple and practical guide showing the conditions under which cold weather is dangerous and when exposed flesh is likely to freeze. Soldiers must be familiar with the wind-chill chart to recognize these dangers.

1-2. EFFECTS OF COLD WEATHER ON MILITARY OPERATIONS

Cold has a numbing effect on skin, which hinders task performance. This must be considered when accomplishing such routine tasks as vehicle maintenance, and making and striking bivouac areas (FM 9-207).

When conditions become extreme and the temperature drops, the problem of survival becomes critical. Under these circumstances soldiers may tend to withdraw emotionally, not wanting to leave the warmth and protection of their shelter or tent. The remedy is physical activity. Try to remain alert and active. Execution of all military duties and the fast and proper performance of group tasks are vital.

COCOON-LIKE EXISTENCE

Many soldiers, when wearing several layers of clothing and with their heads covered, tend to withdraw within themselves and assume a cocoon-like existence. When so clothed, many soldiers' hearing and field of vision are restricted; they tend not to notice their surroundings. Their thinking and reasoning become sluggish.

INDIVIDUAL AND GROUP HIBERNATION

This process is characterized by individuals seeking the comfort of sleeping bags, and by the group remaining in tents or other shelter. Guard and security measures may be neglected, jeopardizing the security and safety of the unit.

1-3. OVERCOMING THE COLD

The body must be protected. It must be kept clean, dry, and warm, with normal body processes maintained. Rest and nourishment are vital. The right approach to cold-weather living will keep you healthy and get you through challenging times. There are four basic rules to follow:

KEEP IN SHAPE

Cold-weather clothing and equipment are heavy, and add weight to your normal equipment load. This hinders movement in snow and uses too much energy. The importance of being in excellent physical condition cannot be overemphasized. Your unit must have a tough, challenging, demanding physical training program.

DRINK PLENTY OF WATER

Water may be hard to get, so you may drink less than you need or only when you are thirsty. This will not give you the water needed to avoid dehydration. Drink plenty of water (at least 3.5 quarts per day) to avoid dehydration and fatigue. DO NOT eat snow as a water substitute; it will lower your body's core temperature.

EAT TO KEEP FIT

Regular, nutritious hot food is needed for top performance. You must eat, even when you are not hungry.

KEEP A GOOD ATTITUDE

You will meet many new challenges, but none that cannot be overcome. Your attitude toward the cold will reflect those of your leaders.

1-4. POSITIVE LEADERSHIP AND THE RIGHT ATTITUDE

Most soldiers, after 3 to 4 weeks of cold-weather training, will be able to cope with the environment. Some soldiers may not be as skilled as others, but most will succeed.

Set an example for other soldiers. Winter conditions can be frightening, trying to meet challenges you have never met before. Any mistake could make you a casualty. Aggressive leadership actions that overcome the challenges of the environment are essential to mission accomplishment.

There are two enemies to overcome in northern operations: the enemy soldier and the cold. The first step in defeating either enemy is having the right attitude. The cold can defeat you psychologically if you are not aware of the symptoms of a poorly prepared unit. The following are suggestions on how to defeat the cold:

- If you become depressed or moody, or do not want to talk, force yourself to get involved with others. Do not leave out the weaker soldier.
- If you find it hard to remember things you have been taught, review orders and drills.
- Remind soldiers that their jobs are to close with, engage, and gut the enemy, and that weapons and equipment must be kept in fighting order. Operations in a northern environment can easily become a camping trip with serious consequences. You must avoid this common ambush.
- Cold weather may slow down the following of an order, but it is not an excuse for not doing it.
- Cold weather may make tasks harder and take longer to do, but it will not make them impossible. With knowledge, proper equipment, and proper training you can overcome the cold and be successful in combat.

CHAPTER 2

Personal Clothing and Equipment

"The freezing German troops were reduced to removing clothes from enemy corpses, improvising strawboots, and taking other emergency measures."

Russia, 1942

2-1. INDIVIDUAL CLOTHING

Soldiers should understand the design principles of the military cold-weather clothing system. These principles are: insulate, layer, and ventilate. There are four ways to make these principles work for you: keep it Clean, avoid Overheating, wear it Loose, and keep it Dry. By remembering the "buzz word" COLD, you can quickly check your clothing.

INSULATE

Insulation material reduces the amount of heat loss. By regulating the amount of insulation, an individual can regulate the amount of heat lost or retained.

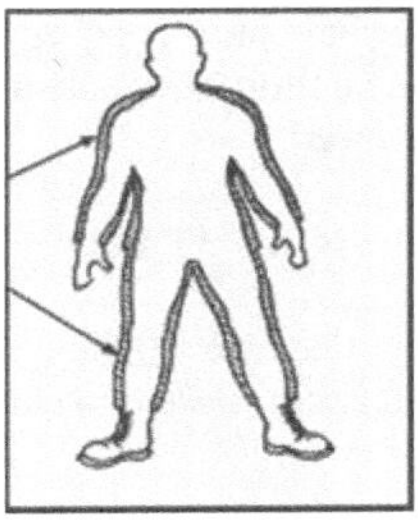

Figure 2-1. Insulation.

LAYER

Several layers of clothing provide more insulation and movement than one heavy garment, even if the heavy garment is as thick as the combined layers. Air pockets trapped between the layers provide insulation and warmth.

VENTILATE

Ventilation helps maintain a comfortable body temperature by allowing excess heat and moisture to escape.

Figure 2-2. C-0-L-D.

KEEP CLOTHING CLEAN

Dirty clothes are cold clothes. Dirt and grease clog the air spaces in clothing and reduce insulation.

AVOID OVERHEATING

Select comfortable clothing and do not overdress. In cold weather, it is better to be a little cool rather than risk being overheated.

WEAR IT LOOSE

All items of the cold-weather uniform are large enough to allow wearing of the appropriate number of layers. Your field jacket may appear too large without all of the layers designed to fit under it. If the uniform items do not fit loosely, the insulation that keeps you warm will be greatly reduced.

KEEP IT DRY

It is important to keep clothing dry, since wet clothing conducts heat away from the body more quickly than dry clothing. Moisture soaks into clothing from two directions: the inside and the outside.

From the outside you must guard against melting snow; from the inside you must guard against sweating. Brush snow and frost from clothing before entering any heated shelter or vehicle, and avoid overheating.

2-2. COLD-WET VERSUS COLD-DRY

As mentioned in Chapter 1, temperatures that are above 14°F (-10° C) are considered cold-wet conditions; temperatures consistently below 14°F will cause cold-dry conditions. Cold-wet conditions usually involve a freeze/thaw cycle, often accompanied by rain or wet snow. This type of weather is hard to dress for, since each soldier must battle the elements to keep his clothing dry. During cold conditions, the ground is usually frozen and the snow is dry. Athough the temperatures during the cold-dry conditions may be lower than during cold-wet conditions, it is easier to stay warm.

COLD-WET COLD-DRY

OR

DOWN TO 14°F BELOW 14°F

Figure 2-3. Cold-wet versus cold-dry.

2-3. THE COLD-WEATHER UNIFORM

There are many layers of clothing that must be worn in a cold weather environment. The first layer of clothing is the T-shirt and shorts. The second layer is a set of thermal underwear, which consists of an undershirt and underpants. You should wear only the clothing needed to keep comfortable and warm.

The scissors-type suspenders must be the last thing next to the outer layer, or else the layers cannot be tucked inside the trousers. The underpants and all succeeding layers of trousers are supported from the metal hooks on the suspenders. The suspenders allow the underpants and trousers to be worn loosely at the waist so that circulation and ventilation are not restricted.

Wear the wool shirt outside the trousers, which provides another method of ventilation. You can increase the waist size of this shirt by removing the stitching in each dart at the sides of the shirt.

Wear the field trousers as an outer garment in the cold-wet environment. In the cold-dry environment, wear the nylon liners inside the field trousers during periods of inactivity. On the inside of each cargo pocket on the trousers is a tape that runs through a hole on the side of the pocket. The tape ties around the thighs and keeps the cargo in the pocket from rubbing and irritating the legs. For warmth and to prevent entry of snow, tie the drawstrings around your ankles outside the boots into the boot lacing, and tuck the bowknots under the trousers.

The field jacket is worn with the liner in both cold-wet and cold-dry areas. Wear the coat and liner directly over the wool shirt. Remove the liner in mild weather or when activity increases.

In a cold-dry environment the parka is worn with a liner. The waist is adjustable for ventilation and warmth with a drawcord, and the skirt has a "split-tail" that can be tied tightly around each leg for warmth. During a period of severe cold, when activity is less, wear the parka with liner over your field jacket with liner. Remember: do not overdress or you will sweat.

Do not dry nylon on drying lines too close to the heat; too much heat will melt the nylon, destroying its insulating value.

Keep the head covered to stay warm; a warm head even warms the feet. Wear the cold-weather cap or balaclava under your helmet or under the hood to your field jacket. When wearing the cold-weather cap as an outer headpiece, the lower flap may be folded up over the top and fastened with hook-pile tape straps in front. In a cold-dry environment, wear the winter hood over the cap or the helmet. Adjust the hood by using the drawstrings on either side of the hood and the elastic strip at the back. To provide minimum protection for the face from the wind, bend the wire in the fur cuff of the hood so that you have a small opening for your face. This allows visibility but less protection.

The feet are hard to keep warm and dry, so protect them. The insulated boots (black for cold-wet conditions and white for cold-dry conditions) are warm and waterproof. The white insulated boot is the best cold-weather footwear today. When you are active, your feet will sweat and, because of the waterproof rubber lining, the moisture will remain in the socks or in the bottom of the boots. The feet, even if damp, will stay warm because the heat given off by your body will remain inside the boots. If the boots are worn for long periods, the feet become white and wrinkled when you remove the boots. This is no cause for alarm. Drying, massaging with foot powder, warming and putting on dry socks will correct this. If you are inactive for long periods, your feet may become cold. Exercises such as knee bends, stamping your feet, running in place, or wiggling your toes will make them warm. Elevate the feet whenever possible to aid blood circulation. Don't wear socks that are too tight, and wear only one pair of cushion-sole socks at a time. Change your socks at least twice a day. Most important, stay active.

Care of your boots is important. Wash boots inside and out with soapy water at least once per month or as required by unit SOP you want your feet to keep warm, you must keep the insulation dry. Inspect your shoes often and repair any holes or punctures immediately. Temporary repairs can be made by patching the hole with any kind of tape; even chewing gum can be used. Seal the hole quickly to prevent the insulation from becoming wet. If one of your boots feels heavier than the other, the insulation is probably wet and they should be turned in for new ones.

Insulated boots must be tested annually to ensure they are waterproof and have less than 0.01 percent moisture in the insulation. The air valve on the outside of the boot is used to equalize pressure when worn at high altitudes and adjusted again upon return to low altitudes. To equalize the pressure in the boot, open the valve for a few seconds and immediately close it. The valve should be kept closed at all other times to prevent moisture from entering. Never open the valve unless the boot pressure around your foot causes discomfort.

You can wear overshoes over the leather boots on wet or muddy terrain. The overshoes are not as warm and dry as the insulated boots. Make sure your socks are dry and that your boots are loosely laced so as not to restrict circulation.

2-4. EXTENDED COLD-WEATHER CLOTHING SYSTEM (ECWCS) (TEST)

The ECWCS is issued as a one-for-one replacement for the current standard cold-wet clothing system. System components use the most recent materials and technology. The result is a reduction in weight and bulk from the current cold-wet clothing system without loss of protection from the cold. The ECWCS uses the layer approach that draws moisture away from the body and should be worn as discussed below.

FIRST LAYER

This consists of polypropylene undershirt and shorts. They are worn next to the skin. They are designed to draw moisture away from the skin, thereby reducing conductive heat loss. You must make sure that no other item is placed between these items and your skin.

SECOND LAYER

The second layer contains a synthetic fiber pile (SFP) shirt and polyester batting trouser liners. The SFP shirt and trouser liners provide insulation. The shirt is designed to dry quickly if it becomes wet and to provide a high degree of insulation. It should not be worn as an exterior garment in wet or windy conditions, since the loose weave provides little protection against wind chill. If the shirt gets wet, it will reduce the insulation value.

The trouser liners are designed to be worn during little or no activity. They should not be worn during heavy activity such as skiing or snowshoeing.

THIRD LAYER

This consists of polyester batting field jacket liner and field trousers. The jacket liner should only be used for extra insulation during periods of little activity. This liner is designed to be used as a separate jacket; it should not be attached to the parka. The trousers provide a durable wind-resistant, water-repellent layer to protect the insulating underpants and trouser liners.

FOURTH LAYER

Windproof, waterproof parka and trousers make up this layer. The parka and trousers provide windproof and waterproof shell garments to protect the first three layers.

The parka should be the outer garment, which is always worn when doing any activity outside a shelter. It is waterproof, yet the fabric is breathable so that body moisture can escape. The hood is designed to give full protection from wind and snow and does not prevent your seeing the area around you. It has underarm zippers to allow air to circulate during heavy activity. The trousers are made of the same material as the parka and are very warm. They should be worn only when necessary to prevent the field trousers from becoming wet.

FIFTH LAYER

This consists of standard camouflage overwhites.

2-5. TIPS ON WEAR AND MAINTENANCE OF CLOTHING

The previous discussion on the layers of cold-wet and cold-dry uniforms describes how each system functions. In some units, the cold-dry field trouser liners are often substituted for the wool field trousers of the cold-wet system and function well in a cold-wet environment. Your commander and unit standing operating procedure (SOP) will determine what uniform and components will be worn and carried in the field.

DRYING WET CLOTHING

Hang items separately from drying lines inside the tent. Do not place anything too close to a stove or over steaming pots; steam will cause further dampness.

DRYING DAMP CLOTHING

Damp items can be hung on the rucksack during movement. Pin damp socks to load-carrying equipment or shirt during movement or just inside the last layer of clothing. Do not place damp clothing in sleeping bags. This will only make sleeping bags wet, and they are harder to dry than clothing.

Small items can be placed between the sleeping bag and the sleeping mat under your legs. When washing clothes without soap, dry-rub and air clothing and sleeping bags.

Never warm your feet next to an open fire. The boots will melt and burn because of their insulation before the feet become warm.

Try never to wear VB boots in a sleeping bag. Sometimes the tactical situation is such that boots must be worn while sleeping. If so, the feet must be dried and massaged with foot powder, and dry socks must be put on. Wearing the VB boot while sleeping does not let your feet dry out. After VB boots have dried, place them underneath your sleeping bag on top of your sleep pad when temperatures are below 0°F.

2-6. COLD-WEATHER EQUIPMENT

The sleeping equipment consists of a sleeping bag, an insulated sleeping pad, and a waterproof bag. There are two types of sleeping bags:

(1) Sleeping bag type I, intermediate cold, temperature range +45°F to +10°F, consists of a mummy-shaped bag with hood, insulated sleeping pad, and a waterproof bag.

(2) Sleeping bag type II, extreme cold, temperature range of +10°F to -50°F, is also available.

Whatever clothing is necessary must be worn with both bags to stay warm and comfortable at the lower temperature ranges. As a minimum, long underwear, sleep shirt, dry socks, and head protection should be worn while sleeping.

The insulated sleeping pad has replaced the pneumatic mattress. Use the insulated sleeping pad under the sleeping bag for warmth and comfort and to keep your sleeping bag dry.

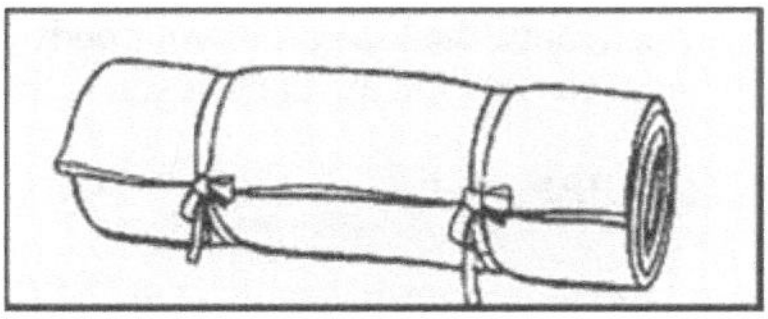

Figure 2-4. Sleeping pad.

The bag must be kept dry if it is to keep you warm. Sleeping bags become wet from outside moisture, from sweating on the inside of the bag, and from water escaping from the mouth or nose into the sleeping bag.

To keep outside moisture from wetting the bag, place protective insulation, pine or spruce boughs, a poncho, or an insulated sleeping pad under the bag. Avoid sweating by wearing the least amount of clothing to keep warm inside the bag, and by using the proper sleeping bag for the temperature conditions. Do not breathe into the bag or moisture will collect and wet the bag or form ice crystals. If your face gets cold, cover it with a towel, muffler, balaclava, or sleeping hood. The sleeping shirt with hood is specifically designed for use with the types I and II sleeping bags.

When it is not in use or being carried, open the sleeping bag wide so that fresh air can get into it. Turn it inside out so that the sun and wind can dry it. Always shake the bag out and fluff it up before and after use to provide for more trapped air between the layers of the wind-resistant/water-resistant outer shell.

Always carry your sleeping bag in the waterproof carrying bag. This protects the bag and helps keep it dry. Ensure that the top bag is securely tied to prevent snow or other moisture from getting inside and wetting the sleeping bag. The following suggestions will help you keep warm and comfortable in the sleeping bag:

- Eat just before getting into the bag. This gives you energy and helps you to stay warm.
- Always relieve yourself just before going to bed. The act of getting out of a warm sleeping bag to go outside can be a chilling experience.

PACKING YOUR RUCKSACK

When you load your rucksack for skiing or climbing, pack heavy items at the bottom and next to the frame. This places the weight on your hips, which is necessary for good balance. Place hard or sharp objects inside the load where they will not rub on the bag and your back. When walking or snowshoeing, raise the center of gravity and take more of the load on your shoulders. Articles that are often needed are placed in the outside pockets for easy reach. Keep maps and other flat objects in the flap pocket.

ADJUSTMENT

Adjust the straps so that you can move the top part of your body and swing your arms freely. Adjust the shoulder straps so that the lower back straps fit just above the belt line.

2-7. LOAD-CARRYING EQUIPMENT

The all-purpose lightweight individual carrying equipment (ALICE) rucksack comes in sizes medium and large. Make sure you are equipped with the large ALICE in a cold environment.

In cold-weather operations, you should follow the packing list set forth in your unit SOP to organize your rucksack. The main consideration is to avoid having to unpack during halts to get frequently needed items. The following illustrates a method of organizing your rucksack:

- **Small external pockets.** These contain small high energy foods (candy, cereal bars) to be eaten while on the move.
- **Large external pockets.** These contain rations to be eaten for morning and evening meals. Take extra socks, scarves, and a spare cap. The openings behind the pockets can be used for skis or other items.
- **External attachment points.** The sleeping pad is attached to the bottom of the pack or under the top flap (wrapped inside sandbags) and other combat items as required.
- **Top flap.** It contains camouflage overwhites and pack cover.
- **Main compartment.** Place the sleeping bag in the bottom of the rucksack. Spare clothes are placed in the upper half of the compartment for easy reach.
- **Camouflage.** A white cotton cover is provided that covers the entire rucksack, excluding the shoulder straps. Use it when wearing your overwhites.
- **Care and maintenance.** Check your rucksack for damages before and after an operation. Repair damages immediately or turn in through supply channels for a replacement.
- **Emergency quick-release of the rucksack.** When quick removal of the rucksack is needed, unsnap the fastener. Pull down on the quick-release fastener on the left shoulder strap, shift the rucksack toward the right shoulder, and let it drop to the ground.

- **Cargo support shelf.** The rucksack can be converted to a cargo carrier by removing the combat pack and replacing it with the cargo support shelf. You can carry square, rectangular, or box-type items such as radios, water cans, and fuel cans.

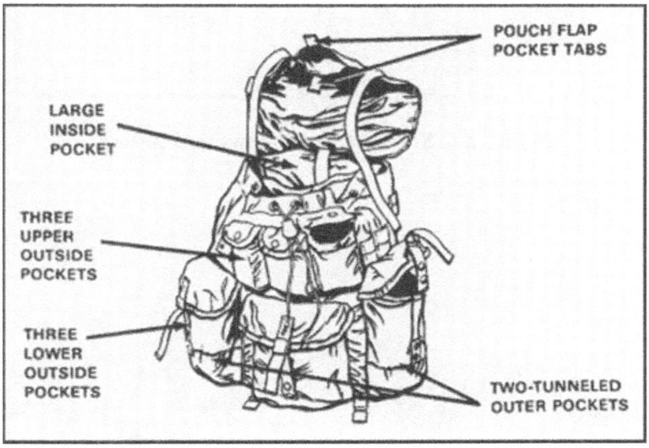

Figure 2-5. Load-carrying equipment.

2-8. OVER-THE-SNOW MOVEMENT EQUIPMENT

Before you are instructed to use snowshoes or skis, your chain command will first determine if movement by foot is possible. In shallow snow (8 inches or less) or heavily crusted snow, movement on foot for inexperienced skiers provides better maneuverability with the least effort.

SNOWSHOES

Snowshoes consist of a white magnesium frame laced with nylon-coated steel cable. They are about 48 inches by 12 inches weigh 3.3 pounds with bindings. When properly adjusted, the toe of the boot will be about halfway across the opening in the snowshoe; the toe strap should be firm, but not tight enough to restrict circulation; and the heel straps should permit some side-to-side movement. When used correctly, snowshoes provide better mobility than walking. In most instances, skis are preferred over snowshoes. When using snowshoes, you cannot back up or swim; be careful when moving over thin ice. The traction needed to ascend or descend with snowshoes can be increased by waxing. The use of ski poles with snowshoes aids movement and reduces fatigue.

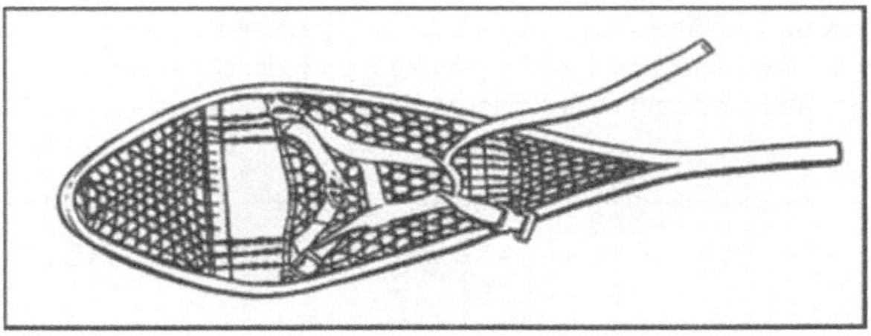

Figure 2-6. Snowshoe and snowshoe binding.

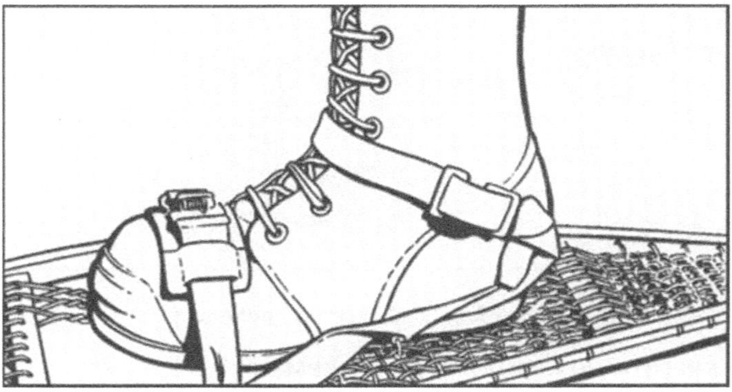

Figure 2-7. Snowshoe binding attached to VB boot.

SKIS

The standard military skis are about 208 centimeters (82 inches) in length, and have a hole in the tip for towing, a groove in the heel for attaching mohair climbers, and metal edges.

SKI POLES

Ski poles can be very useful in steep terrain. The military ski pole consists of a steel shaft with rubber handgrips and baskets, and a leather wrist strap. The poles come in three lengths: 130 centimeters (51 inches), 137 centimeters (54 inches), and 147 centimeters (58 inches). When fitting the poles, you should stand on the floor and place the pole at your side; it should come to the point of your shoulder (about 1 inch below the top of the shoulder).

MOHAIR SKI CLIMBERS

The mohair climbers are attached to the bottom of the skis for added grip when climbing steep slopes. The climbers consist of a cotton-backed mohair pile and are attached to the skis by five web straps.

2-9. MISCELLANEOUS EQUIPMENT

SUNGLASSES

Polarized sunglasses should always be worn on bright days when the ground is covered by snow. Failure to wear the sunglasses, even on overcast days when the need seems less, may result in blindness.

PROTECTIVE MASK

There is an M4 cold-weather winterization kit available for your protective mask. Make sure you install this winterization kit for operations in the northern environment.

CANTEENS

In cold weather the plastic canteen will freeze very quickly if carried in the standard fabric carrier. Since the canteen cannot be placed directly on heat, there is no quick method for thawing. Whenever possible, the canteen should be carried in one of the interior uniform pockets, or wrapped in clothing and placed in the rucksack to ride against your back. In extreme cold, do not fill the canteen over two-thirds full; this allows for expansion if the water freezes.

The insulated canteen is a one-quart capacity stainless steel vacuum bottle. It is provided with a cotton duck cover and a stainless steel cup. A new arctic canteen will keep water from freezing at -40°F for 6 hours. The following are some tips in using the arctic canteen:

- Fill the canteen using your canteen cup.
- Fill to within half an inch of the neck.
- Do not drop or dent.
- Put it in your sleeping bag at night.
- Carry the canteen in an M60 bandoleer near your body.

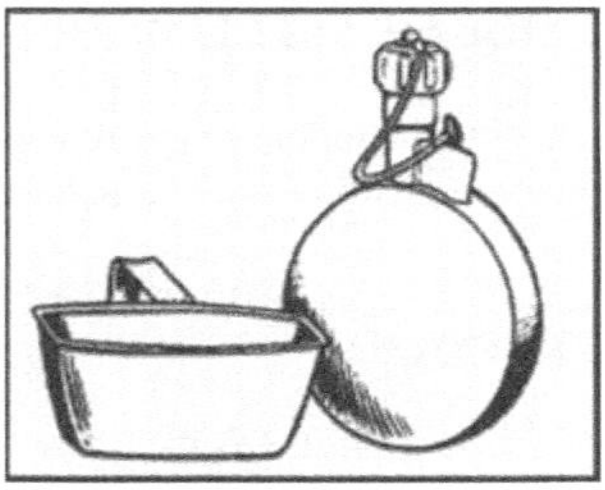

Figure 2-8. Insulated canteen with cup.

CHAPTER 3

Tents and Heating Equipment

"A well-sheltered defender enjoyed a marked advantage over his exposed attacker..."

Ironside, 1918

3-1. GENERAL

To conduct successful military operations in cold weather and maintain a high level of combat efficiency, you must have heated shelters. Tents and stoves are a vital part of cold-weather equipment.

3-2. TENT GROUP EQUIPMENT

The following is a list of the tent or ahkio group for an infantry squad:

1 Ten-man tent with liner
1 Yukon stove
1 Metal or plywood stove base
2 Five-gallon gasoline cans (one with white gasoline per platoon)
1 Five-gallon water can (filled)
2 Cases, meals, ready-to-eat
1 Climbing rope (120 feet)
1 Box of candles
2 Squad cook sets
2 One-burner squad stoves
2 Axes
2 Saws
1 Hatchet with sheath
1 D-handle shovel
1 Sledgehammer
1 Pick
1 Tripod for gas can (set)
3 Pairs of snowshoe bindings (spare)
1 Lantern (extra mantles)
1 Box of soda bicarbonate (fire extinguisher)
1 Roll of soft wire
5 Plastic bags
3 Rolls of toilet paper

3-3. AHKIO

The ahkio is the squad's method of transporting the tent group equipment in the winter. It is a 200-pound capacity fiberglass sled that can be used for transporting tent group equipment, for carrying weapons and ammunition, as a firing platform, and for the evacuation of casualties.

3-4. PACKING

When packing the ahkio, the load must be evenly distributed. Place heavy equipment on the bottom and slightly to the rear of center. Loading lighter equipment toward the top will prevent the ahkio from becoming top heavy. Tools such as shovels, axes, and saws are packed at the sides for easy access in trailbreaking; however, caution must be taken to ensure that these items do not stick out. The canvas cover is folded over the load. To keep the load from shifting and to prevent snow from getting under the cover, lash the load tightly by crisscrossing the lashing rope from the lashing rings on one side of the sled to the rings on the other side. An additional ahkio, issued on a basis of one per two 10-man tents, should be used to carry camouflage nets and support systems. In very wet or dry snow, the movement of an ahkio is easier when wax is applied to the running surface.

3-5. TEN-MAN ARCTIC TENT

The 10-man arctic tent with liner is a six-sided pryamidal tent supported by a telescopic pole. The tent is designed to accommodate 10 men and their clothing and equipment. It will hold additional personnel by leaving packs and equipment outside the tent, or by lowering the center pole, which allows the sidewalls to spread out to cover more ground surface. The tent has two doors, which permits joining of tents for access from one tent to another. A snow cloth is attached to the bottom of the sidewalls for sealing the tent to the ground. The snow cloth should always be to the outside of the tent with pine or spruce boughs underneath, or it may freeze to the ground. This allows easy escape in case of an emergency if the doors are blocked. The tent is ventilated by four built-in ventilators on opposite sides and near the top of the tent.

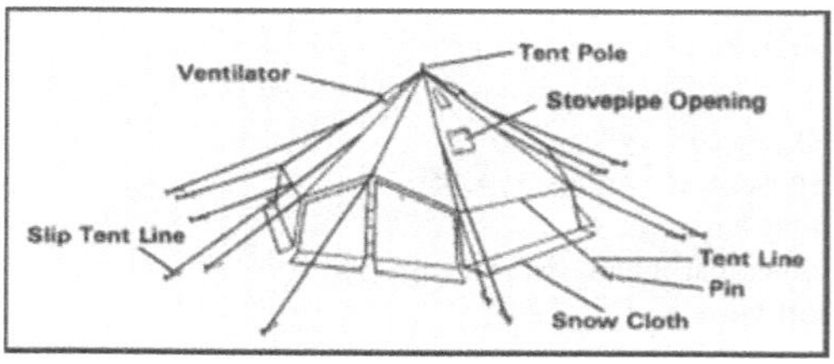

Figure 3-1. Ten-man arctic tent.

3-6. BUILDING ARCTIC TENTS

The shelter is easily built by four trained soldiers. Consider the following when pitching or striking tents in snow or on frozen ground.

Snow should be cleared to the ground surface to obtain a lower silhouette and gain advantage of ground temperatures, which are generally warmer than upper air temperatures. Enough pine or spruce boughs should be placed on the tent and floor, and under the snow covers, for insulation and comfort. If an insulating material is not used, the heat of the stove will thaw the ground insulation in a sea of mud. The snow covers will freeze to the ground, preventing escape from the tent in case of an emergency. When it is impractical to remove the snow to the ground level, an adequate tent site can be prepared by packing the snow with skis or snowshoes until a firm base is provided for pitching. The center pole is placed on a log or other suitable support to keep it from sinking into the snow.

In open terrain, with a strong wind, it will be necessary to build a 1-meter-high snow wall on the windward side of the tent to protect it from the wind. The tent is pitched with the entrance angled about 45 degrees from the downwind side. When the tent is set up, the snow cloth around the bottom of the tent should be spread out so that snow or logs can be placed on it to help in anchoring the tent.

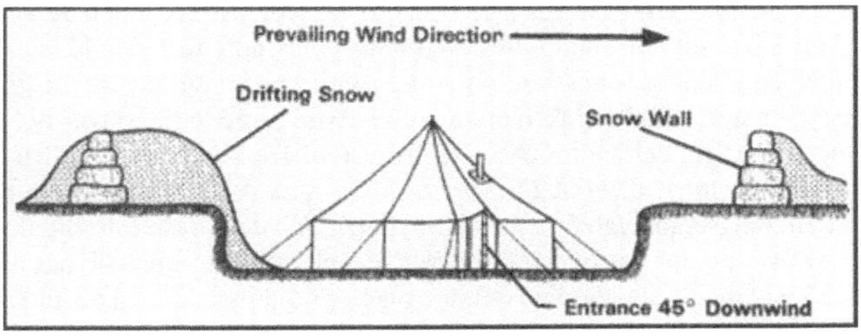

Figure 3-2. Windbreak.

Driving tent pins into frozen ground is impossible. Tents may be pitched and anchored by attaching the tent lines to trees, branches, logs, or stumps. If these are not available, holes are dug into the snow for the purpose of using "deadmen." This is done by digging a hole into the snow large enough to insert a pole or log about 1 meter long with the tent line attached. The hole is filled with snow and well packed. The packed snow freezes, securely anchoring the tent. In powder or dry snow, water would be poured into the hole.

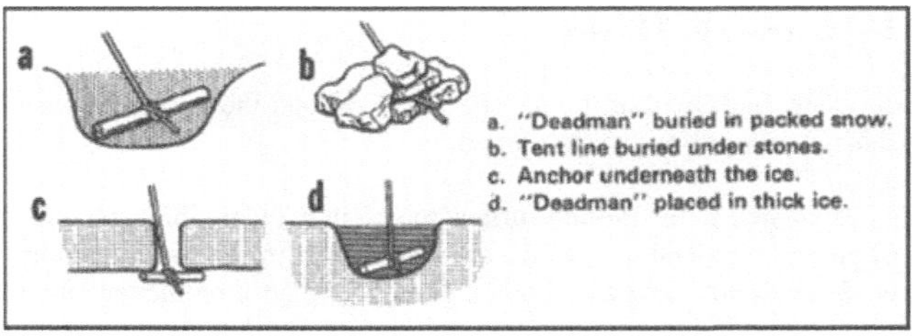

Figure 3-3. Deadmen.

The smaller tent (five-man) provides the commander more flexibility in distribution of his forces. You can never be assigned a tactical mission in the arctic without provisions for a shelter. This tent enables a smaller unit to be assigned an independent mission, such as patrol or reconnaissance mission. It is also six-sided, pyramidal, and supported by a telescopic tent pole. It is designed to accommodate four to five men and their clothing and equipment. The tent can provide shelter for a rifle squad or other similar unit when rucksacks are placed outside the tent.

3-7. INSTRUCTIONS FOR PITCHING TENTS

Instructions for pitching tents are sewn securely to the tent just inside the door. Although the instructions are easily read and followed, there is one thing that should be remembered when pitching either tent. After the tent is spread out flat on the ground, the six corner eave lines should be staked out and loosely secured to the stakes before the tent is raised. These lines are yellow for easy identification. The stakes or tie-down points for these lines should be located about 2 meters from the tent. After this is done, proceed with the tent pitching as indicated in the instructions. The door eave line should be propped up by placing the line over a pole (tree branch or other object higher than the door entrance) and then stake the line out to a pike or tie-down point. This procedure keeps the door from sagging and makes the zipper work better. Drying lines should be attached to the liner inside the tent to allow drying of clothes.

In winter the tent will normally be covered with ice and snow, which must be removed before the tent is struck. This can be done by shaking the tent or by beating it with a mitten or stick. If the snow cloth is frozen to the ground, the snow and ice around it must be removed by chopping or shoveling. This must be done carefully to avoid damage to the snow cloth. One way to do this is to ease the shovel between the cloth and the ground and gently pry the cloth away from the ice.

When the bottom of the tent is properly sealed and the door is zipped, moisture will form on the inside of the tent and collect on clothing and equipment. Carbon monoxide and fumes from the stove can accumulate to a dangerous degree. To avoid this, ensure that the ventilators near the peak of each tent are open.

The roof of the shelter must be steep enough to shed snowfall. With both the 5-man and 10-man tent, the area that will become the floor of the tent is dug, packed, and insulated with pine, spruce boughs, or similar material. Snow walls should then be built to protect against wind.

3-8. YUKON STOVE

The Yukon stove is used to heat the 10-man, 5-man, and GP small tent. In addition to providing heat, the top surface of the stove and the area beneath the stove may be used to heat rations and water. The stoves use standard MOGAS but may also be operated with diesel fuel or JP-4 without modification. NEVER MIX FUELS. When solid fuels (wood, coal, etc.) are used, the stove must be modified by removing the burner unit, closing the burner opening, and turning over the wire grate so there is space below for draft and ashes.

WARNING

DO NOT USE A SPARK ARRESTER ON THE M1950 YUKON STOVE.

A flue cap, issued as a component part of the Yukon stove, shields the stovepipe top from the wind and prevents the wind from blowing smoke or gases down the pipe into the stove and tent. The flue cap is equipped with three 4 1/2-meter lines that are used to steady the stovepipe and anchor it in high winds. These guy lines must be anchored to the tent or tent ropes, not to the ground or nearby trees. The stovepipe hole cover must be rolled up and tied off to prevent nylon ties from touching the stovepipe.

A firm base for the stove must be provided to keep it from melting down into the snow. The base should be 35 centimeters by 70 centimeters and be made of lightweight metal or salvaged plywood. If plywood is used, it should be covered with metal or aluminum foil. If those materials are not available, a base can be made out of small logs about 10 centimeters in diameter.

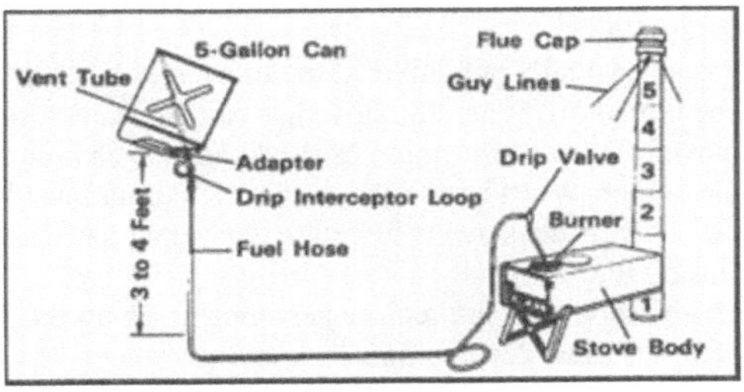

Figure 3-4. Yukon stove.

A simple stand for holding the fuel can is built by obtaining three poles 2 meters long. The poles are tied about two thirds of the way up and then spread out to form a tripod. The fuel can should be at least 1 meter higher than the stove. If the fuel can is wobbly or if there is a wind, the can should be tied to this tripod. Make sure the can is tilted so that the air is trapped in the uppermost corner.

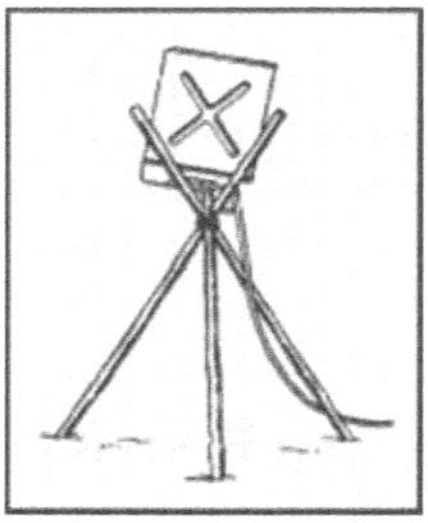

Figure 3-5. Tripod for gas can.

3-9. PRECAUTIONS

The following precautions must be observed when the Yukon stove is being used:

- All stovepipe connections must be tight, and the tent stovepipe shield properly adjusted. The stove must be level to ensure that the burner assembly will spread the flame evenly throughout the stove.
- Ensure that the fuel line is made of natural rubber and suitable for extreme-cold operation. It must be protected so it cannot be accidentally pulled loose and must have a drip intercept loop to prevent fuel from running down the line and into the tent.
- Rate of fuel flow must be checked regularly. It will change with the fuel level and will need adjustment. All fuel supplies must be kept outside the tent.
- The stove should never be left unattended when it is burning. Allowing the stove to become too hot may cause the stove body to become overheated and warped and burn the tent down.
- If the flame is accidentally put out, or if the fuel can is being changed, the drip valve must be turned off. When the stove has cooled, excess fuel inside the stove must be removed. Two or three minutes must pass so gas fumes escape before relighting the burner. When burning gasoline, the burner must be cool before relighting the stove. If the stove is lit before the burner has cooled, the fuel will vaporize, causing it to explode.
- If the tactical situation does not require personnel to be awake 24 hours a day, turn the stove off during sleeping time.

The following are precautions when **burning wood or coal**:

- Fuel should be fed in small amounts until there is a hot bed of coals.
- The stove should not become overheated.
- Gasoline should not be poured on the wood or coal inside the stove to hurry starting the fire.
- Ashes should be removed regularly and not allowed to accumulate higher than the bottom of the grate.
- NEVER MIX FUELS.

3-10. SQUAD STOVE, M1950

The one-burner squad stove is a cooking and heating unit for two to five men. It is used when operating in an isolated or forward area when separated from the ahkio group. The stove is small, compact, and lightweight, and it will operate on either MOGAS or white gasoline. The initial pressure for its operation is generated by a few strokes on the hand pump. Thereafter, the pressure is generated by heat from the burner evaporating the fuel in the stove tank.

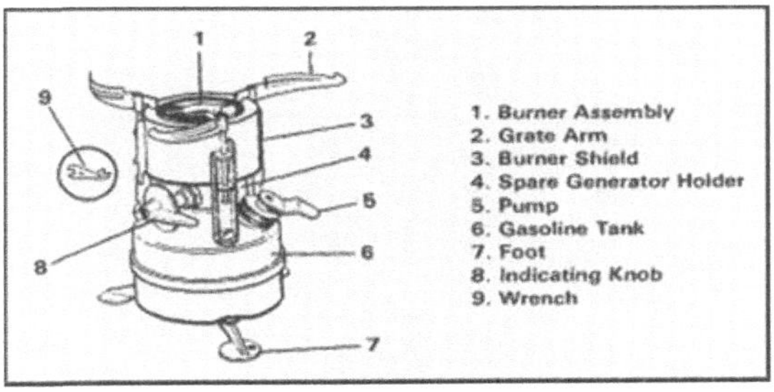

Figure 3-6. Gasoline cooking stove, M1950.

To operate the squad stove, do the following:

- Snap out the three feet and set stove on a level surface.
- Raise the grate arms and turn them outward so that the inner tabs can be engaged in the slots in the burner shield.
- Close the valve by turning the indicating knob clockwise to the OFF position.
- Unscrew the pump caps and remove the pump assembly.
- Fill the gasoline tank three-fourths full.
- Wipe off any gasoline that may have been spilled during filling and replace the pump cap.
- Pump 10 strokes to build up initial pressure.
- Turn the indicating knob to the ON position for 3 seconds. This allows the gasoline to rise through the generator to the burner and preheated cup.

- Turn the indicating knob to the OFF position to prevent excess fuel from accumulating in the burner assembly.
- Light the wick and allow the flame to burn for about 3 minutes or until the flame turns blue.
- Turn the indicating knob to the LIGHT position. Have a second match ready to relight the burner immediately if the flame goes out.
- When the flame in the burner turns blue, turn the indicating knob to the ON position.
- Pump an additional 10 strokes to maintain pressure within the tank. The stove is now ready for use.
- If the flame burns unevenly or dies during operation, turn the valve to the OFF and ON position several times, relight quickly, and pump several additional strokes.
- To turn the stove off, turn the indicating knob to the OFF position and allow the flame to go out.

CHAPTER 4

Small Arms and Ammunition

"Rifles and machine guns and even the breechblocks of artillery became absolutely rigid ... and steel parts cracked."

General Rendalic, Eastern Front, 1942

4-1. GENERAL

Cold weather will greatly affect the operation of your weapon. You must be aware of these effects to ensure employment of your weapon under poor weather conditions.

4-2. CARE, CLEANING, AND MAINTENANCE

Your weapon will work properly under extreme cold conditions if you take care of it. Normal temperate zone lubricants thicken in cold weather, and stoppages or sluggish actions of your weapon will result. During the winter, disassemble your weapon completely, clean it thoroughly, and remove all lubricants. Oil the entire weapon with a light coating of cleaner lubricant preservative (CLP).

CAUTION
DO NOT USE CLP ON A MORTAR OR ANY COMPONENT PART OF A MORTAR.

Figure 4-1. Weapons will 'sweat" when taken in to a heated shelter.

A main problem is to keep snow and ice, which may disable your weapon, out of the working parts, sights, and barrel. When the weapon is not in use, keep it out of the snow either in an improvised weapon stand or in a tripod arrangement. If this is not possible, stand it up with the butt end in the snow.

Condensation ("sweating") will form on the weapon when you take it from the cold into a heated shelter. The condensation will continue for about 1 hour, after which you should clean the weapon. If you don't, the condensation will freeze when you take the weapon outside, and it may not fire. The weapon must be kept as near the floor as possible to lessen condensation. If parts of the weapon do become frozen, warm them and move the working parts slowly until they are unfrozen. If you can't warm your weapon, remove all ice and snow and move the parts gently until action is restored. The use of CLP will prevent malfunctions by eliminating condensation.

When firing, do not let the hot parts of your weapon come in contact with the snow. The snow will melt and, as the weapon cools, will form ice. When changing barrels, don't lay them on the snow.

To prevent bipods and tripods from sinking into the snow, improvise for added flotation. This is done by placing them on snowshoes, ski poles, branches, ahkios, etc. Baseplates should be on solid ground or sandbags to prevent sinking into the snow and damage to the weapon.

Figure 4-2. Bipods and tripods will sink in the snow.

AMMUNITION

Small arms ammunition should be kept at the same temperature as your weapon. Ammunition clips and magazines must be cleaned of all oil and preservative and checked often. All ice, snow, and condensation must be removed. Ammunition boxes must be kept closed in order to keep out snow and ice.

Stored ammunition should be kept in packed boxes, raised off the surface by placing them on boughs, cardboard, or other similar material, and covered to protect them from the snow. Ammunition should be well marked for easy location and identification in case it becomes covered with snow.

Ammunition resupply may become restricted. Everyone must be aware of the necessity for ammunition economy and fire discipline. Loaded clips, magazines, or single rounds dropped in the snow become quickly lost; therefore, careful handling of ammunition is essential. Never waste your ammunition. Make every round count.

When contact with the enemy is imminent, troop compartments and aircraft should be kept as close to freezing as possible to avoid condensation. However, many munitions have limitations at lower temperatures, which must be recognized in training.

CHAPTER 5

Rations and Diet

> ***"I ate my boots and a pair of pants ... tea and seal-skin gloves for dinner."***
>
> ***Greely, Arctic Expedition, 1883***

5-1. GENERAL

Most of what we eat and drink is used in maintaining our body heat, while only a small portion is used in producing energy for physical work. You must ensure adequate caloric intake in cold-weather operations. About 4,000 calories per day are necessary for personnel performing physically demanding work in the cold. Efficiency may drop rapidly if this level is not maintained.

The body loses liquid at a very fast rate in arctic conditions, regardless of how carefully you adjust and ventilate your clothing. The exertion of movement on foot, preparation of bivouac sites and defenses in the snow, etc., take a toll in sweat and loss of moisture in the breath. These liquids must be regularly replaced, preferably by hot drinks, which provide extra calories if they contain sugar.

5-2. RATIONS

Rations provide the needed calories to live and fight effectively. When eaten in their entirety, rations contain the right amount of carbohydrates, fats, protein, and vitamins. The proper intake of these essential items depends on the entire ration being eaten in properly spaced meals. The lack of concern regarding eating caused by the cold, combined with the difficulties and inconvenience of cooking, may tempt soldiers to miss meals. The principles of sound leadership and discipline in cold weather require that meals be prepared and that the entire ration be eaten, warmed when possible. Save snacks (cookies, chocolate bars, etc.) for between meals and when on the march.

5-3. LIQUIDS

Because there is a large amount of fluid loss in the cold, you should drink a minimum of 3.5 quarts of water a day when heavy physical activity is involved. Water is usually available either from streams or lakes or by melting snow or ice. A limiting factor may be fuel needed to melt the snow or ice. To save time and fuel, water should be obtained from running streams or a lake. The milky water of glacial streams must be allowed to stand until the sediment settles. When a hole is cut into the ice to get water, it should be covered by a snow block or something similar to slow refreezing.

Ensure that as much of the daily liquid intake as possible is hot drinks, such as soups and cocoa. Your local command may request a special food allowance of warming and cooling beverages. Main meals should begin with soup, and between-meal snacks should include a hot drink.

When running water is not available, ice or snow must be melted. Ice produces more water in less time than snow. When melting snow, first put a small amount into the cooking pot; then more can be added. Continue this process until there is enough water. It must be purified by boiling rapidly for 15 minutes.

When heating water, use all available stoves, since this is a most time-consuming operation. Melting and boiling enough snow for a drink may take 30 to 40 minutes. Areas designated for ice or snow that is to be used for water must be sited far upwind of the latrine and garbage disposal sites. Before going to bed, ensure that stoves are filled and enough water has been prepared for breakfast. Do not let other soldiers eat snow or ice; doing so lowers the body's core temperature.

Never drink alcohol on operations or when exposed to low temperatures. The aftereffects of alcohol can lead to a feeling of false security. You may forget the main rules to prevent cold injury, like wearing your gloves or not going outside without shoes to relieve yourself.

CHAPTER 6

Hygiene and First Aid

"...the first frostbite cases have made their appearance. We amputated two feet and will amputate many more."

Amosoff

6-1. GENERAL

The care of your body needs special attention in cold weather. Ensuring cleanliness in the field, particularly in the cold, is important. It is easy to let yourself go for long periods without bathing, but if you wish to be part of an effective combat force, you must keep clean.

6-2. HYGIENE TIPS FOR THE SOLDIER

Face, hands, armpits, and crotch should be washed daily. If water is not available, take a "bath" using cornstarch or rub your body briskly with a dry towel. The cornstarch removes excess oil and perspiration from skin and hair.

Shave regularly -- daily if water is available or every 2 to 3 days if water is in short supply. Shaving just before sleeping will allow the face maximum recovery time before going out into the cold, reducing the possibility of superficial frostbite.

Teeth should be cleaned daily. If a toothbrush is not available, a clean piece of cloth wrapped around a finger or a chewed twig may be used as a substitute.

Underwear should be changed at least twice a week. If it is not possible to wash your underwear, crumple it, shake it, and air it for at least 2 hours.

Ensure that all eating and cooking utensils are cleaned after each use.

In the dry arctic you will find the need to oil your unprotected and exposed skin once every week (with an oil such as suntan oil).

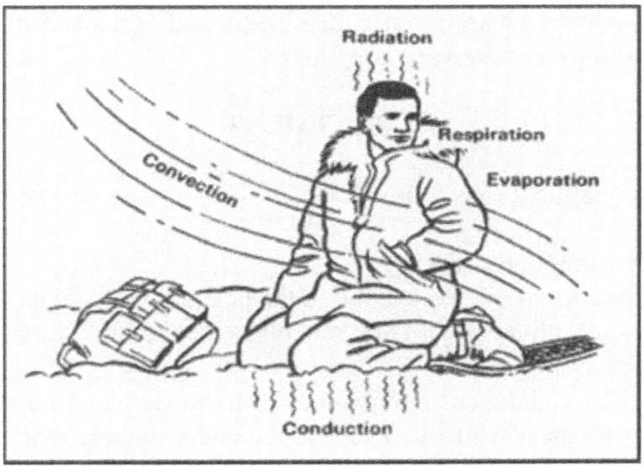

Figure 6-1. Mechanisms of heat loss from the body.

6-3. HYPOTHERMIA

Hypothermia is a lowering of the temperature of the body's inner core. This happens when the body loses heat faster than it can produce it. You must be able to recognize the symptoms of hypothermia and to treat the condition immediately.

PREVENTION

To prevent hypothermia, adhere to the following:

- Stay physically fit.
- Keep active.
- Use the uniform properly and keep it dry.
- Eat properly and often.
- Drink plenty of liquids, at least 3.5 quarts per day, when performing physically demanding work.
- Be prepared for and know how to deal with quick changes in the weather.
- Bivouac early before judgment is weakened by fatigue.

SYMPTOMS

Be aware of the following symptoms:

- Shallow breathing or absence of breathing.
- Faint or undetectable pulse.
- Victim is first cold, then stops shivering.
- Loss of feeling.
- Poor coordination; victim has trouble walking.
- Impaired speech.

- Confusion, withdrawn appearance, depressed, and/or uncoordinated. An uncaring attitude and/or glassy stare are also clues.

TREATMENT

The recommended treatment is as follows:

- Prevent any further heat loss.
- Get the victim out of the wind and into the best shelter available.
- Replace any wet clothing with dry, and place the victim in a sleeping bag, if one is available.
- Place as much insulation as possible between the victim and the ground.
- Add heat by the best available means to the victim's neck, groin, and sides of chest. This can be done by using a hot water bottle, Yukon stove, campfire, or your own body heat.
- Provide the victim with something warm and nutritious to drink if he is conscious. Calories may be added by using sugar in hot, sweet drinks.
- DO NOT massage the victim.
- DO NOT give alcohol to the victim.
- Evacuate the victim to the nearest medical treatment facility as soon as possible. This condition is life threatening until normal body temperature is restored.

6-4. FROSTBITE

Frostbite results when tissues freeze from exposure to temperatures below 32°F. The degree of injury depends upon the wind-chill factor, length of exposure time, and proper protection. Persons with a history of cold injury are prone to frostbite. There may be a tingling, stinging, an aching sensation, or a cramping pain. The skin first turns red and then becomes pale gray or waxy white. Frostbite can be classified as superficial or deep.

WIND CHILL

The combination of wind and low temperatures creates a condition known as wind chill. For example, with the wind calm and a temperature of -20°F there is little danger of wind chill. If the temperature is -20°F and there is a wind of 20 knots, the equivalent chill temperature is -75°F!

Under these conditions there is GREAT DANGER and your exposed skin can freeze within 30 seconds. You also create your own wind when you walk, run, or ski. As examples, walking creates a wind speed of about 5 mph and running 10 mph. Wind is also created by moving vehicles, aircraft propellers, and rotors. The wind-chill chart is provided to show how wind and temperature combine to create wind chill; use it. Cold injury can disable you like an enemy bullet.

WIND SPEED		COOLING POWER OF WIND EXPRESSED AS "EQUIVALENT CHILL TEMPERATURE"																				
KNOTS	MPH	TEMPERATURE (°F)																				
CALM	CALM	40	35	30	25	20	15	10	5	0	-5	-10	-15	-20	-25	-30	-35	-40	-45	-50	-55	-60
		EQUIVALENT CHILL TEMPERATURE																				
3-6	5	35	30	25	20	15	10	5	0	-5	-10	-15	-20	-25	-30	-35	-40	-45	-50	-55	-60	-70
7-10	10	30	20	15	10	5	0	-10	-15	-20	-25	-35	-40	-45	-50	-60	-65	-70	-75	-80	-90	-95
11-15	15	25	15	10	0	-5	-10	-20	-25	-30	-40	-45	-50	-60	-65	-70	-80	-85	-90	-100	-105	-110
16-19	20	20	10	5	0	-10	-15	-25	-30	-35	-45	-50	-60	-65	-75	-80	-85	-95	-100	-110	-115	-120
20-23	25	15	10	0	-5	-15	-20	-30	-35	-45	-50	-60	-65	-75	-80	-90	-95	-105	-110	-120	-125	-135
24-28	30	10	5	0	-10	-20	-25	-30	-40	-50	-55	-65	-70	-80	-85	-95	-100	-110	-115	-125	-130	-140
29-32	35	10	5	-5	-10	-20	-30	-35	-40	-50	-60	-65	-75	-80	-90	-100	-105	-115	-120	-130	-135	-145
33-36	40	10	0	-5	-15	-20	-30	-35	-45	-55	-60	-70	-75	-85	-95	-100	-110	-115	-125	-130	-140	-150
WINDS ABOVE 40 HAVE LITTLE ADDITIONAL EFFECT		LITTLE DANGER					INCREASING DANGER (Flesh may freeze within 1 minute)						GREAT DANGER (Flesh may freeze within 30 secs)									

Figure 6-2. Wind-chill chart.

PREVENTION

Frostbite is preventable if the individual soldier takes care of himself. It seldom occurs in people who are maintaining enough body heat. It is most commonly associated with an overall body heat loss resulting from poor equipment, reduced food intake, dehydration (water loss), overtiring, injury, or a combination of these factors. The feet, hands, ears, and exposed areas of the face are most easily hurt by frostbite and must receive constant care. The following are rules to help prevent frostbite:

- Do not wear tight boots and socks.
- Use the "buddy system" to check exposed areas, especially when wind is present.
- Carry extra socks and mitten liners.
- Do not stay still for long periods.
- Use caution when cold and wind are combined.
- Check feet during halts.
- Ensure that you eat well and have hot drinks as often as possible.

SYMPTOMS OF SUPERFICIAL FROSTBITE

Redness occurs, followed by powdery flaking of the skin. Affected areas of dark-skinned soldiers may appear dull and grayish. Blister formation occurs 24 to 30 hours after exposure, followed by flaking of superficial skin.

SYMPTOMS OF DEEP FROSTBITE

Common symptoms are as follows:

- Loss of feeling occurs in the affected area, leaving it pale yellow and waxy looking.
- When the frostbitten area thaws, it is painful.
- Frozen tissue may feel solid or "wooden" to the touch.
- When exposed to inside temperatures, blisters may appear in 12 to 36 hours.

- Discoloration (red-violet) appears 1 to 5 days after the injury.
- Gangrene usually results.

TREATMENT FOR FROSTBITE

Determine whether the frostbite is superficial or deep. If the exposure time was short, the frostbite will probably be superficial. If the exposure time was long, the frostbite will probably be deep.

The following is a recommended treatment for frostbite:

- Move the casualty to a warm and sheltered area.
- Do not rub the frostbitten area with snow or apply cold water soaks.
- Do not rewarm the affected area by massaging or exposing it to open fire.
- Rewarm the face, nose, or ears by placing your hands on the frozen area. Rewarm frostbitten hands by placing them under clothing and against the body.
- Close the clothing to prevent further loss of body heat.
- Rewarm the feet by removing the boots and socks. Place the bare feet under the clothing and against the abdomen of a buddy. Once the feet are warmed, put on dry socks and boots, if available. If the victim must wear the wet socks and boots, he should exercise his feet by wiggling his toes.
- Loosen tight clothing and remove jewelry.
- Improve circulation by exercise.
- Do not allow the victim to use alcohol or tobacco. Alcohol increases loss of body heat; tobacco causes narrowing of blood vessels in the arms and legs.
- Reassure the victim.
- Protect frozen tissue from further cold or trauma, and evacuate the casualty to the nearest medical treatment facility.

Deep frostbite is most common and harmful to the feet and less common to the hands and ears. When deep frostbite does occur, the victim should be moved to a sheltered area and evacuated to a medical treatment facility immediately. If possible, do not allow the victim to walk if his feet are frozen and avoid treating or thawing the affected area. Thawing of deep frostbite in the field increase pain and invites infection, greater damage, and gangrene. Do not try to thaw deep frostbite; let the physician do it.

6-5. DEHYDRATION

Dehydration can cause serious physical problems and make you more susceptible to other problems, such as frostbite and hypothermia.

PREVENTION

The following are suggestions to prevent dehydration:

- The **minimum** daily liquid requirement for personnel performing demanding physical work in the cold is 3.5 quarts per day.
- By the time you feel thirsty you are already dehydrated. Drink whenever possible, particularly during halts.
- Ensure that canteens are full before any type of movement.
- During periods when water is scarce, do not drink coffee in large quantities. It will pass through you quickly. Water should be the first choice.
- Check your urine spots in the snow. Dark yellow or brown indicates dehydration.

SYMPTOMS

Common symptoms are as follows:

- Red indicates a severe condition needing immediate medical evacuation to the nearest medical treatment facility.
- Lack of appetite.
- Dry mouth, tongue, and throat.
- Stomach cramps and/or vomiting.
- Headaches.

TREATMENT

Recommended treatment is as follows:

- Keep the victim warm.
- Give him plenty of liquids.
- Make him rest.

6-6. CARBON MONOXIDE POISONING

Whenever a stove, fire, gasoline heater, or internal combustion engine is used indoors, there is danger of carbon monoxide poisoning. Fresh air in living and working quarters is vital. Carbon monoxide is a deadly, odorless gas.

PREVENTION

The following are suggestions to prevent carbon monoxide poisoning:

- Use stoves and lanterns in well-ventilated areas and tents.
- Ensure that stoves and lanterns are functioning properly.
- Do not let personnel warm themselves by engine exhaust.
- Always have windows cracked in vehicles with a heater in use.

- Use a tent guard or shut the stove off when sleeping.

SYMPTOMS

Common symptoms are as follows:

- Headache, dizziness, confusion, yawning, weariness, nausea, and ringing in the ears.
- Bright red color on lips and skin.
- Victim may become drowsy and collapse suddenly.
- If personnel are found unconscious in an enclosed shelter, carbon monoxide poisoning should be suspected.

TREATMENT

Adhere to the following for proper treatment:

- Move the victim to open air.
- Keep the victim still and warm.
- If the victim is not breathing, administer mouth-to-mouth resuscitation.
- Administer cardiopulmonary resuscitation, if necessary.
- Immediately evacuate the victim to the nearest medical treatment facility.

6-7. SNOW BLINDNESS

Snow blindness is caused by ultraviolet and ultrablue rays of the sun being reflected from a snow-covered surface into the eyes. This condition can occur even in cloudy weather. In fact, it is more likely to occur in hazy, cloudy weather than in sunny weather.

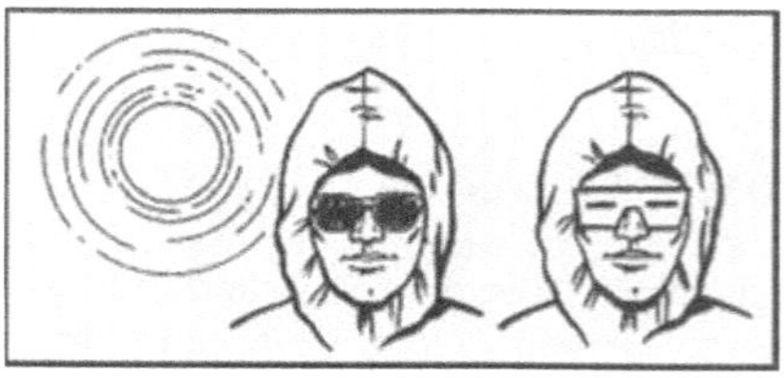

Figure 6-3. Snow blindness.

PREVENTION

Prevention is easy; wear your issued sunglasses. If sunglasses are not available, make slitted glasses from cardboard, thin wood, tree bark, or similar material.

SYMPTOMS

Common symptoms are as follows:

- A scratchy, sandy feeling under your eyelids.
- Redness and watering of the eyes.
- A headache.

TREATMENT

Adhere to the following for proper treatment:

- Blindfold the patient using a dark cloth.
- Reassure the patient.
- Evacuate the patient to a medical treatment facility.

6-8. SUNBURN

Sunburn is a serious hazard because of the reflective qualities of the snow, especially at higher elevations.

PREVENTION

Require soldiers to use a sunscreen and lip balm.

SYMPTOMS

Redness of skin with slight swelling is one symptom of sunburn. Prolonged exposure to the sun may cause pain and blistering and, in severe cases, chills, fever, and headaches.

TREATMENT

Soothing skin creams may be helpful if the swelling is not severe.

6-9. TENT EYE

Tent eye is an inflammation of the eye caused by fumes from stoves and lanterns used in a poorly ventilated tent or shelter. It is prevented by proper ventilation of tents or shelters. The treatment is fresh air.

6-10. TRENCH FOOT/IMMERSION FOOT

Trench foot and immersion foot are injuries caused by the prolonged exposure of skin to cold or wet conditions. In extreme cases, the skin dies and amputation of the foot or leg may be necessary.

PREVENTION

The following helps prevent trench foot or immersion foot:

- Change to dry socks at least twice daily.
- Dry and massage your feet regularly with foot powder.
- Wear VB boots and change socks regularly.
- If leather boots are worn, dry them whenever possible.

SYMPTOMS

In early stages, the feet and toes are cold, numb, and still, and walking becomes challenging. The feet swell and become painful.

TREATMENT

In early stages, keep the feet dry, clean, and exposed to the air. In later stages, evacuate the victim to a medical facility.

6-11. CONSTIPATION

At very cold temperatures, personnel tend to put off the natural urge to have a bowel movement, resulting in constipation. This can be a painful casualty producer.

PREVENTION

Eat fresh or canned fruit and eat regularly. Drink plenty of liquids. Try to have a daily bowel movement at your normal time.

SYMPTOMS

Some common symptoms are stomach cramps, dizziness, and headaches.

TREATMENT

Prevention is the best treatment. If symptoms persist, get medical help.

6-12. DIARRHEA

Diarrhea can be treated by drinking tea leaves with water.

6-13. HEAT CRAMPS

Heat cramps are caused by an excessive loss of salt in the body. This loss of salt causes muscles to spasm, and they are unable to relax. After prolonged physical activity in a cold-wet environment, the signs and symptoms of heat cramps could appear in the arms, legs, and/or stomach. It is possible to experience heat cramps in cold-weather environment when undergoing extreme physical exertion.

SYMPTOMS

The soldier experiences muscle cramps in his arms, legs, and/or stomach. He is pale and has wet skin and is experiencing dizziness and extreme thirst.

TREATMENT

Have the soldier drink one fourth of a canteen of water. Give the soldier one canteen of water with one packet of salt stirred in. Have him drink the canteen of salted water over a 30-minute period. If cramping of muscles continues, evacuate the soldier to the nearest medical treatment facility.

NOTE

Do not give the saltwater to a soldier who has an upset stomach. Have him drink a canten of unsalted water.

6-14. HEAT EXHAUSTION

Heat exhaustion is caused by dehydration and loss of body salt from extreme physical activity in a cold-wet environment. It is possible to experience heat exhaustion in a cold-weather environment when undergoing extreme physical exertion.

SYMPTOMS

The soldier may feel dizzy, weak, and/or faint. The skin feels cool and moist to the touch. He may feel nauseated or have a headache.

TREATMENT

The following is a recommended treatment for heat exhaustion:

- Have the soldier drink one canteen of water to relieve the symptoms.
- If he complains of cramps, give him one canteen of water with one packet of salt stirred in. Have him drink the salt solution over a 30-minute period.

NOTE

Do not administer salt if the soldier has an upset stomach. Have him drink a canteen of unsalted water.

- Loosen all tight-fitting clothes.
- Elevate the feet above the level of the heart.
- If the soldier is unable to drink water due to an upset stomach or if the symptoms have not improved within 20 minutes, evacuate the soldier to the nearest medical treatment facility.

CHAPTER 7

Bivouac Routine

"Had we lived, I should have a tale to tell of the hardihood, the endurance and the courage of my companions which would have stirred the hearts of every Englishman. These rough notes and our dead bodies must tell the story."

CPT Robert Scott

7-1. GENERAL

The essential requirements for survival and mission accomplishment in the cold/arctic environment are warmth, food, and shelter. As covered in Chapter 3, you are provided with tents that can be erected quickly. Tents can be lost or destroyed and separated from the ahkio groups in the confusion of combat. When this occurs, timber and snow, which are usually available, will be used for shelters. A detailed explanation of the bivouac routine should be contained in the unit SOP.

7-2. BIVOUAC SITE

Bivouac sites, other than those that are permanent nature in rear areas, either are tactical or are designated to provide living accommodations within a defended position. The tactical bivouac is usually not occupied for more than one night, is established at or after last light, and needs little preparation. It is organized for a rapid and disciplined response to enemy action. The defensive position is likely to be occupied for longer periods and must be established accordingly. The layout and placement of tents and shelters must be coordinated within the defense and in accordance with the unit SOP.

7-3. SITE SELECTION

The selection of bivouac areas is important and needs careful consideration. The challenge of area selection varies with the tactical situation, weather conditions, and terrain. The bivouac area will be tactically located in accordance with the principles of security and defense.

There must be concealment from air and ground observation and wind. Forested areas pose few challenges in comparison to areas that are located above the tree line. In selecting areas, ensure that local camouflage materials are available for use with the snow camouflage nets. Areas that offer better bivouac sites than others are explained below.

FORESTED AREAS

Forests provide excellent sites. Camouflage is easy and building materials, firewood, and wood for insulation are readily available. Although forests provide the best bivouac site, temperatures there will normally be colder than those in open areas.

MARSHY GROUND

In winter, good sites are frozen swampy areas, the banks of streams, and the shores of lakes. Some swampy areas may not freeze because of warm springs or marsh gases. Deep valleys should be avoided since colder air gathers there.

OPEN AREAS

Avoid open areas because of strong winds, drifting snow, and poor concealment. If avoidance is not possible, then pitch tents where they can be sheltered by natural windbreaks. These are normally found in depressions or behind ridges of ice on lakes. In areas where natural wind falls do not exist, build snow walls to provide wind protection. In areas with high winds, snow gathers on the lee side, making it necessary to clear the side and tops of the tents, often to prevent the weight of the drifting snow from collapsing the tent.

MOUNTAINOUS AREAS

Mountainous terrain is characterized by strong winds, cold, and a general lack of concealment above the timberline. The wind over head creates extensive lees near cliffs and steep rock faces. Avalanches can occur on steep slopes; avoid their likely path. Signs to look for are:

- Heavy snow overlaying a crust of ice (usually the result of a freeze at the end of a period of thaw).
- New snow on steep slopes if the temperature is rising.
- Evidence of avalanches in previous years; an area cut through the trees (branches stripped off, and/or broken tree stumps at different heights) or a smooth clear strip of ground down the mountainside.

Don't Pitch Tents Where Snow is Likely to Drift

Figure 7-1. Location of tent.

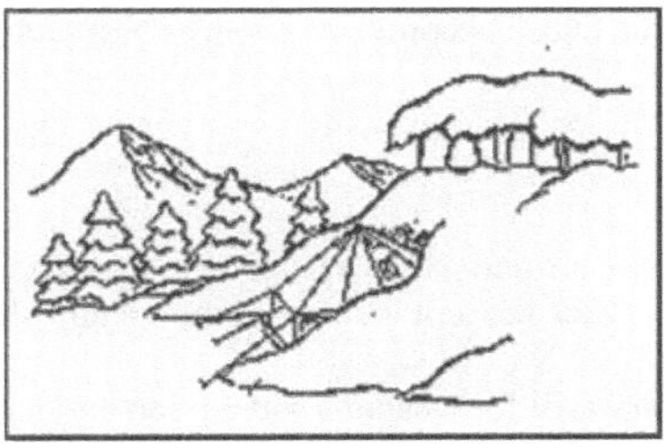

Figure 7-1. Location of tent (continued).

When contact with the enemy is likely, the unit SOP will identify specific considerations when selecting the site for a tactical bivouac. For example:

- The location of the enemy.
- Enemy patrol habits.
- The air threat.
- Dominating features that identify enemy avenues of approach.
- The wind direction that can carry sounds of movement (the bivouac site should be downwind of likely enemy avenues of approach, especially when wind speeds are low).
- Availability of natural camouflage.
- Amount of moonlight and weather conditions. In good weather or bright moonlight, the bivouac site must be in an area of deep shadow (in the middle of the woods or on the northern slope of a ridge) or the tents will be silhouetted.

7-4. RECONNAISSANCE OF BIVOUAC SITE

The main body should be preceded by a reconnaissance party whose task is to lay out the bivouac site. They should:

- Identify the bivouac site and a false position.
- Establish a trail plan.
- Select defensive positions.
- Mark the location of tent sites.
- Identify areas for cutting boughs.
- Act as guides for the main body on its arrival.
- Provide security during occupation by the main body.

7-5. TRAIL PLAN

Trail plans must be established before entering the bivouac site. Normally, the incoming trail will be extended well beyond the area of the bivouac and may lead to a false position. Leading off the incoming trail establish another trail, usually at an extreme angle to entry. The trail network that leads to each subordinate unit area, and within, must follow the inside of the tree line. This prevents enemy observation of trails from the air or ground.

7-6. OCCUPATION

Guides from the reconnaissance party will arrange to meet the main body and identify the layout of the bivouac site. Once the unit is in position, the unit should:

- Identify a temporary area for weapons and equipment during establishment of the site.
- Identify the exact location and type of defensive positions to be built.
- Determine exact tent sites regarding defensive positions, natural shelter, and camouflage.
- Identify sleeping areas and type of shelters to be built when tents are not used.

7-7. BOUGH-CUTTING AREAS

Bough-cutting areas for bedding or improvised shelters should be for all personnel in the unit. These areas must be where activity is not likely — well behind the unit's defensive area or in a decoy bivouac site.

Cut boughs from the bottom portion of trees above the snow and leave the trees standing. Be sure to cut only a few boughs from each tree. A tree stripped of its limbs is easily observed from the ground and from the air.

Cut firewood near the bough-cutting area to allow the same trail to be used. Dead spruce, pine, or squaw wood makes the best firewood. Standing deadwood burns the best and hardwoods the longest. In an arctic and subarctic environment, true hardwoods are rare. Of the trees available, birch, tamarack, and spruce are the hardest woods. Burn the upper parts of dead trees during daylight because they give off lighter colored smoke. The lower parts of the trees contain resin and tar, which burn well but make more and darker smoke. In the daylight this will allow your position to be easily identified by air and ground observation.

7-8. GARBAGE

All garbage should be double bagged and transported to the rear or disposed of in accordance with the unit SOP. Units should never leave evidence of their presence. Carry all trash, if necessary, until it can be disposed of.

7-9. LATRINE

Normally there will be a central latrine within the bivouac site, if the site is not too widely spread. One latrine can usually serve the needs of a platoon-size element. It must be placed downwind of the bivouac site, but not so far that soldiers won't use it. The area identified can be as elaborate as a pit or cross-cut-type latrine that is made windproof by branches, snowblocks, and ponchos. It should be properly camouflaged. It can also be an area where plastic bags are located. Double-bag, tie, and transport all waste to the rear for disposal as identified in the unit SOP. Site urinal areas around a "dedicated tree" or the gasoline tripod, which will hide the discolored snow and will be easily spread about when leaving the bivouac site.

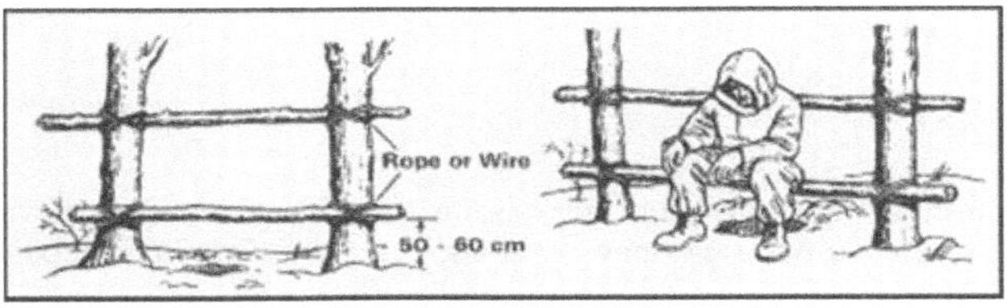

Figure 7-2. Cross-tree type latrine.

7-10. WATER POINTS AND SNOW AREAS

When obtaining water for drinking purposes by melting snow, set aside and restrict an area for this purpose only. The site must be upwind from the bivouac area and isolated from the latrine and garbage areas. The snow must be clean, white, and free of fungus from trees. NEVER use snow that is discolored. Remember — disinfect the water using purification tablets and bring it to a rolling boil for a minimum of 15 minutes.

7-11. STORAGE

Storage problems in the winter are increased by snow, low temperatures, thaws, and limited storage space. Space in any shelter is limited. Store only those items inside the tent that are affected by the cold, or must be immediately available. The remaining equipment should be centralized, well marked, and covered. Store ammunition and fuel separately. Place wood or metal containers on boughs or poles to keep the containers from freezing to the ground. The unit SOP will specify what equipment is placed inside or outside of the tent and where.

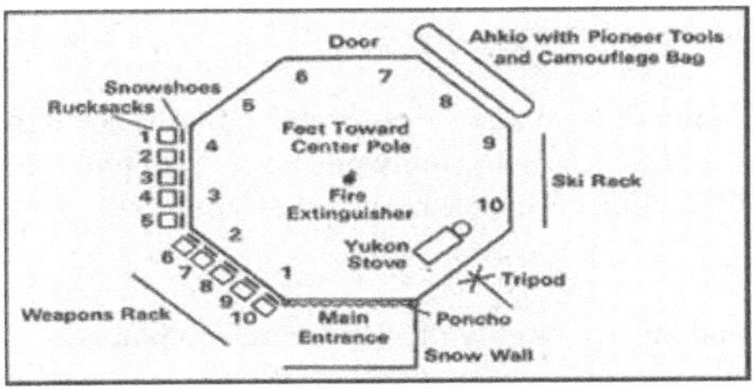

Figure 7-3. Ten-man tent.

CHAPTER 8

Individual Movement

"Anticipate every situation which can arise in battle and think out your reaction to it well in ADVANCE."

Hoote

8-1. GENERAL

Snow affects movement many ways. Heavy snow cover hinders or slows movement, either cross-country or on roads. Snow cover also blankets many terrain features and hides obstacles.

The effect of snow on movement varies greatly with its depth and physical characteristics. Men on foot cannot travel well over flat terrain or roads when the snow is deeper than 30 centimeters. However, movement can be improved by the use of skis or snowshoes.

SKIS	SNOWSHOES
• Faster • Less Work • Harder to Learn	• Slower • More Work • Easy to Learn

Figure 8-1. Using skis or snowshoes.

8-2. SKIS

Skis provide greater speed in moving, particularly over cleared trails. Troops mounted on skis and towed behind small unit support vehicles, or similar vehicles (ski-joring), are effective means for rapid cross-country movement. In heavy vegetation, skiing will always be more physically taxing than snowshoeing.

8-3. SNOWSHOES

Learning how to use snowshoes requires less training. Snowshoe movement is more practical in confined areas, such as assembly positions, field trains, and mortar and artillery positions. If traveling on foot, you should become skilled on both skis and snowshoes.

Your ability to use, and the availability of, over-snow equipment is a major factor in learning to travel well over snow-covered terrain. With previous training in cross-country skiing and if equipment is available, travel on skis is recommended. In most snow conditions and over most types of terrain, skis provide the fastest and most energy-saving form of travel. Use of snowshoes requires little training, but your speed will be slower and travel more exhausting.

Snowshoes are **more** exhausting than skis. However, a light crust on the surface of the snow prevents skis from sinking and provides for fast and easy skiing. A crust hard enough to support a man makes travel on foot feasible, but even then travel on skis is recommended.

In a survival situation, improvise equipment for travel if snow is loose and deep. Make snowshoes of willow or other springy wood using a wood separator and thong, wire, cord, etc.

USE SKIS WHEN:

- Speed is essential.
- Long distance must be traveled.
- Stealth is necessary.
- Conditions allow.

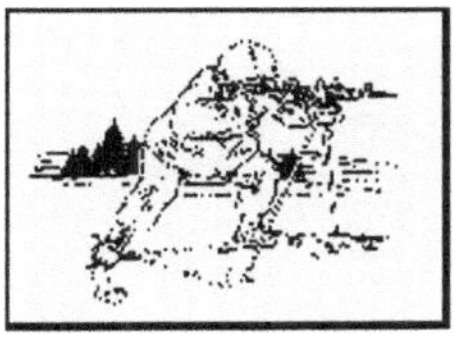

USE SNOW SHOES WHEN:

- Moving through heavy brush.
- Speed is not essential.
- Troops are not proficient on skis.

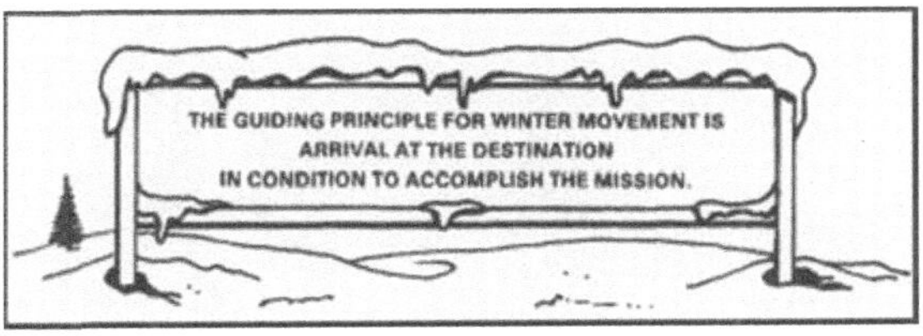

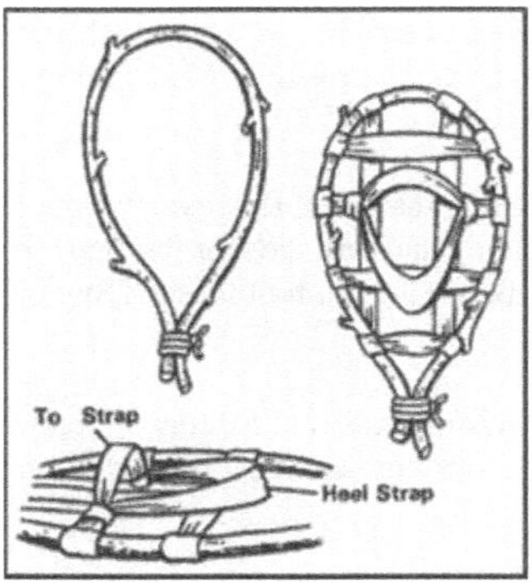

Figure 8-2. Improvised snowshoes.

CHAPTER 9

Land Navigation

"No - I've never been lost; but for about three weeks I was kinda puzzled."
Brigadier General Fergusson

9-1. GENERAL

In some cold-weather environments, accurate up-to-date maps may not be available and landmarks may be few or nonexistent. When the terrain is snow covered, one piece of ground looks like the other. Whatever the circumstances, you must be able to find your way — if not by map reading, by some other means of navigation.

9-2. CHALLENGE OF LAND NAVIGATION

Only poor road nets may exist in some areas. These will be of little use in cross-country navigation. If large-scale maps exist, they may be unreliable.

Long nights, falling or blowing snow, and fog limit visibility in winter. Tracks and landmarks may be erased by heavy snowfall. Accurately identify landmarks, through both ground and air observation, to prevent mistakes.

Conditions of daylight are different from those found in temperate regions (moderate climate). In winter there are long nights and short days.

Magnetic disturbances may be common, thereby making compass readings unreliable. Magnetic declinations may be extreme and can vary considerably at close points.

Many small lakes, ponds, and creeks will be found that are not on the map. The size and location of waterways may vary from year to year and from season to season and cannot be relied on to pinpoint a position. Large lakes and rivers can be used as important aids to navigation and movement.

9-4. DISTANCE AND DIRECTION

When traveling from one point to another, it is necessary for you to know the distance between them beforehand, then plot your journey accordingly.

9-5. MEASURING DISTANCES

PACING

In temperate zones, pacing is the simplest method of measuring distance. In cold-weather areas, snow cover and the use of skis and snowshoes make pacing challenging, and it must be practiced repeatedly.

USE OF FIELD WIRE

The best method for accurate ground measurement is for two men to use a piece of salvage field wire 50 meters long. The lead man marches off in the desired direction trailing the wire behind him. The second man signals by jerking the wire when the first has gone 50 meters and then moves forward to the first man's position. The first man marks the spot and continues forward for another 50 meters. This is repeated until the desired distance is reached.

ESTIMATION

If no other method of measurement is available, the length of time spent in steady progress along a route may be used to estimate the distance traveled. The accuracy of this method depends on how well you know your marching speed over all types of terrain on foot, skis, or snowshoes.

9-6. DETERMINING DIRECTION

COMPASS

The use of the compass is the most common way to determine direction. The magnetic field becomes weaker as you get nearer to the magnetic poles and can be easily affected by other objects. It is important to make sure that the compass is not affected by iron fragments in the ground, vehicles, or metal objects on you.

THE SUN

There are two ways to determine direction by the sun:

- At noon, in the northern hemisphere, shadows fall to the north of objects; in the southern hemisphere, to the south of objects.
- Estimate direction by using a watch and the sun. In northern latitudes the hour hand is pointed toward the sun. An imaginary line halfway between the hour hand and 1200 will point south. In southern latitudes the procedure is different. Then point the 1200 toward the sun. An imaginary line halfway between the hour hand and 1200 will point north.

NATURE

Nature will help in determining direction as in the following samples:

- Single evergreen trees, growing away from other trees, will always be more bushy on the south side.
- The bark of birch and poplar trees is whitest on the south side and darkest on the north.
- You should know from which direction the prevailing winds blow. Trees will be bent in the direction the wind is blowing or expected to blow.

THE STARS

The North Star may be used to determine true north. This star can be easily found by its position relative to the Big Dipper. In latitudes under 70 degrees when traveling north, the North Star makes a good steering mark since its bearing is usually only 1 degree from true north and is never more than 2 1/2 degrees away. In higher latitudes above 70 degrees, the North Star is too high in the sky to indicate good direction.

A star near the horizon can be used with a bearing within 2 degrees of your compass course. However, it will only be good for up to 30 minutes when you are traveling north and only 15 minutes when traveling south. After these intervals, select another star.

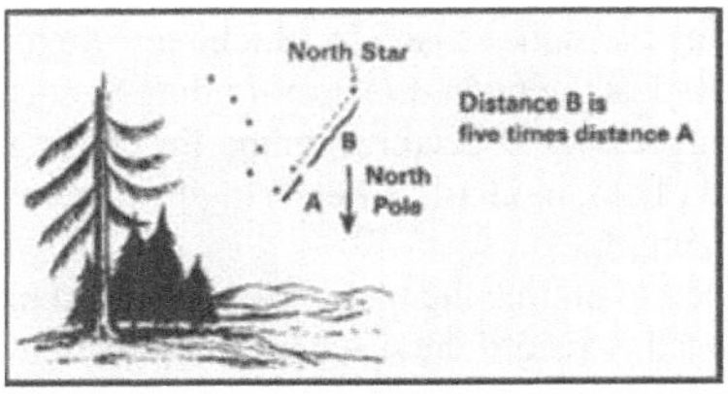

Figure 9-1. Determining direction using North Star.

South of the equator, the constellation Southern Cross will help you locate the general direction of south and any other direction. This group of four bright stars is shaped like a cross that is tilted to one side. The two stars forming the long axis, or stem, of the cross are called pointers. From the foot of the cross, extend the stem five times its length to an imaginary point. This point is in the general direction of south. From this point, look straight down to the horizon and select a steering point.

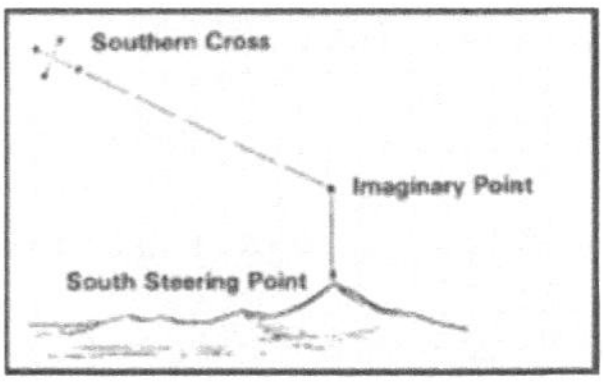

Figure 9-2. Determining direction using Southern Cross.

9-7. NAVIGATION BY DEAD RECKONING

Dead reckoning consists of plotting and recording a series of courses from a known point, each measured for distance and direction. These courses lead from the starting point to the destination. This lets the navigator determine his position at any time, either by following his plot or by comparing his actual position on the ground to his plotted course. All the details are previously estimated and recorded.

EQUIPMENT FOR DEAD RECKONING

- Maps are used for choosing the route and for plotting the route taken as the march progresses.
- Aerial photographs are usually used in addition to maps. They must be vertical photos taken during the same season in which they are to be used.
- The compass is used as the main direction-finding instrument.
- The protractor is necessary in dead reckoning for plotting direction and distance on the map, aerial photo, or chart. When using it, remember to convert magnetic azimuth to grid azimuth.
- A route card is used to outline the plan of the proposed march. Keeping a log is mandatory. It is used to record the distance and direction.
- The log is a written record of where the soldier has started, the distance and direction traveled, and his position at any time in relation to the planned route, destination, and starting point.

THE DRILL OF DEAD RECKONING

Know the starting point and destination and, having examined the map or aerial photo for the best route, prepare a route card. This describes each leg of the proposed route in terms of distance and direction. As the march continues, keep a careful record of each bearing taken and the distance, and record this in the log. There may be a change to the planned route because of terrain or enemy action. Therefore, the necessary changes are made using the map or aerial photo as a guide, and they are recorded in the log.

9-8. HINTS FOR THE NAVIGATOR

The navigator must keep the compass warm to speed up recording his position. When no steering marks exist to the front, he should march on a back azimuth. A steering mark may be a natural or man-made structure to the rear. If the steering mark is not clear, he must keep an eye on it after taking an azimuth so that it doesn't become lost. Never take magnetic azimuths in the area of metal objects.

When visibility is poor, only close-in steering marks will be available. Under these conditions try to pick up steering marks along the correct azimuth. This can be done correctly only when the route to each steering mark follows a straight line. Take compass checks often to ensure that the correct azimuth is being followed. Set the compass for a night course. A steady unshifting wind will also aid in keeping direction.

Depending on the area of the world, other signs can be used. For example, in the barren and semibarren lands of eastern Canada, snowdrifts point to the northwest and windblown trees to the southeast because of prevailing winds. In Scandinavia, scratches on rocks and cliffs caused by the glacial period (ice age) run northwest to southeast.

CHAPTER 10

Survival

"Each day the number of my dogs dwindled and the sleds were broken up to cook those animals that we ate ourselves."

Peary, Arctic Expedition, 1906

10-1. GENERAL

Your mission may take you to an arctic or subarctic region. You expect to have all personal equipment and other members of your unit with you wherever you go. But you could find yourself alone in a remote area with little or no personal gear.

10-2. ACTION WHEN LOST

Brief the soldier on the route before setting out on a march; he should memorize details if possible. A soldier can become separated from the main body of troops, especially if an action is fought. This applies mainly to night battle or bad weather. It can also happen if he is careless and march discipline is poor.

WHEN LOST WITHIN A KNOWN LOCALITY

If all is quiet and there are no battle sounds or aircraft to guide you to friendly lines, go back to the last known position. If you cannot go back, consider carefully the journey already made and the distances and directions from the starting point.

Opinions should be taken from the group. Search parties should be sent out to find the next known steering mark. Trails must be marked carefully so that the search party can return to guide the main group forward or to rejoin the group, should their search be unproductive. Meanwhile, the rest of the group should seek shelter.

If the search party fails to find any known location, perform the steps below, but remember to think clearly and keep calm.

CONDUCT WHEN LOST

When you are certain that you are lost:

- Stay together. Meet your emergency as a group; don't wander off as individuals; never allow stragglers.
- Take shelter. If you have a tent, set it up and get in it. If you do not have a tent, build one of the improvised shelters. The type of shelter is not important as long as you protect yourself from the weather.

- Keep warm. If you are lost in a forest, light a fire (depending on the enemy situation). If you are lost in an area where there are no trees, do not use your gasoline stove to heat your shelter. Use your stove only to prepare hot food and drinks. Conserve your fuel supply; you won't get any more. Get into your sleeping bag to stay warm.
- Check food supply. All the food and water in the group should be collected and rationed systematically. Supplement the food supply by shooting large animals and trapping or snaring small ones.
- Check communications. If you have a radio set, use it to contact your parent unit for help. If you are in a forest, prepare fires to attract search planes. Smoke is better by day and flame at night. Mirrors, ground signals, etc., are helpful in attracting search parties. Arrange necessary ground/air signals for search planes.
- Rest. The more you rest the less energy you expend and the less food you need.
- Adhere to survival tips. Carry safety matches inside the butt plate of your rifle or in your pockets. Carry candles, which are easy to light, provide heat and light, and help in starting a fire. If you have no matches, fire may be produced by striking the flint on your waterproof match box with the blade of a knife and collecting sparks in dry tinder. Dry moss and bark make good tinder to start your fire. Fires should be built in layers with sticks being laid out like the spokes of a wheel.
- Carry a two-cell flashlight. It can be seen from the air or on the ground as far as line-of-sight will permit.
- Ground/air signals can be made with small fires to form the identification characters (see Appendix C).
- Survival depends on clear thinking and resourcefulness. The main things to remember are to keep calm, think, try to help each other, keep together, and keep warm.

10-3. IMPROVISED SHELTERS

Sometimes tents or other regular shelters are not available. In summer, if the weather is mild, you may need protection only from insects. In the winter, however, you cannot stay in the open for long periods unless you are moving. In case of emergency, you must know how to protect yourself from the effects of the weather.

If natural shelters such as caves or rock ledges are available, they should be used. If they are not, a temporary improvised shelter must be built.

The type of shelter to be built depends on the equipment and materials available. By the proper use of materials available, some sort of shelter can be built during any season of the year. In open terrain a shelter can be built using ponchos, canvas, snow blocks, or other materials. Snow caves, snow trenches, snow houses, or snow holes may be built in the winter. In the woods, a lean-to is normally preferred to other types of shelter. In cold-weather areas, nature provides you with the means to prepare a shelter. Your comfort, however, greatly depends on your initiative and skill at improvising.

PONCHO SHELTERS

The simplest type of shelter can be made by pulling the poncho over your sleeping bag. For additional comfort, various types of shelters and lean-tos can be made by attaching your poncho to trees, tree branches, or poles.

To build a one-man shelter from one poncho, spread the poncho on the ground, hood side up. The hood opening must be tightly closed by adjusting and tying the drawstrings. The poncho is raised in the middle of its short dimension to form a ridge, with the stake out at the corners and sides. Snow, sod, or branches are used to seal the side and one end of the shelter to provide additional protection from the wind and to retain heat.

Figure 10-1. One-man shelter.

A two man shelter can be made from two ponchos. Spread the ponchos on the ground with the hood side up and the long sides together so that the snap fastener studs of one poncho may be snapped into those of the other poncho, The ponchos are raised where they are joined to form a ridge. The shelter is then staked out at the sides and corners. A third poncho may be snapped into the other ponchos to form a ground cloth.

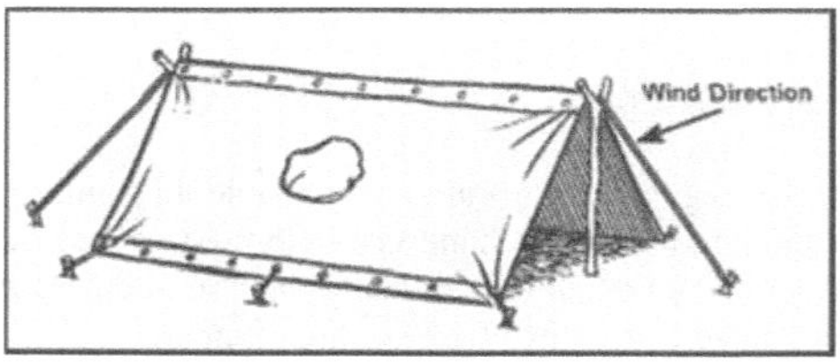

Figure 10-2. Two-man shelter.

LEAN-TO SHELTER

The lean-to shelter is built in forested areas and is made of trees and tree limbs. A poncho, a piece of canvas, or a parachute, in addition to boughs, may be used for covering. Depending on the number of people to be sheltered, two types of lean-tos (single and double) are built. Boughs should be laid in the shelter like overlapping shingles (the same as on a live tree).

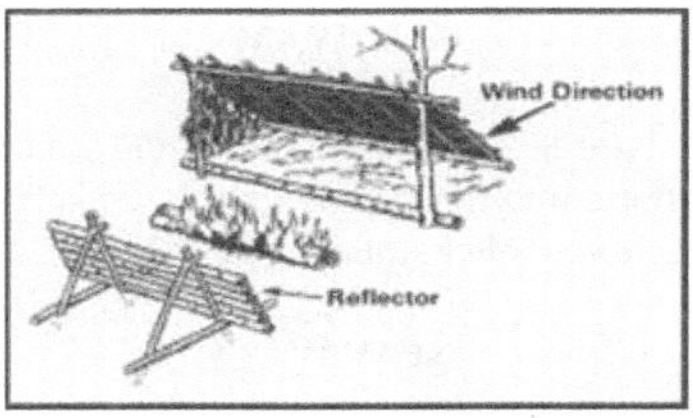

Figure 10-3. Single lean-to.

To save time and energy, use two trees the right distance apart and sturdy enough to support the crosspiece. If suitable trees are not available, use two forked poles or build two A-frames to hold the crosspiece. A large log is laid at the back for added height. Stringers about 3 meters long and 10 centimeters in diameter are then placed 45 centimeters apart from the crosspiece over the top of the log in the rear. Available coverings, ponchos, boughs, and so forth are then placed on top of the stringers. A double lean-to is made by building two single lean-tos facing each other with the fire in between.

Any kind of an open fire can be used to heat the lean-to. The best kind for a large lean-to is the log fire. Precautions must be taken to prevent the fire from burning too hot and burning down the shelter. A fire reflector may be built of green logs or poles to reflect the heat into the shelter and to serve as a windbreak to keep the fire burning steadily.

TREE SHELTERS

In wooded areas, the tree-pit shelter furnishes temporary protection. Select a large tree with thick lower branches and surrounded with deep snow. Enlarge the natural pit around the tree. The walls and floor are lined with branches and boughs. Try not to disturb the snow that may be on the branches.

An emergency shelter for one man can be built by cutting down a coniferous tree at a point about 1 meter from the ground.

Figure 10-4. Tree-pit shelter.

After the tree has fallen, the branches on the underside are trimmed away and the cut material used for insulation on the ground.

WIGWAM

A wigwam or tepee can be built in wooded areas by tying a number of poles near the top and spreading them at the bottom to form a large circle. This framework is then covered with ponchos, a parachute, or any other suitable material.

SNOW WALL

In open terrain with snow and ice, a snow wall may be built for protection from strong winds. Blocks of compact snow and ice are used to build a windbreak. Be careful not to get covered up by drifting snow, which may accumulate on the lee side of the wall.

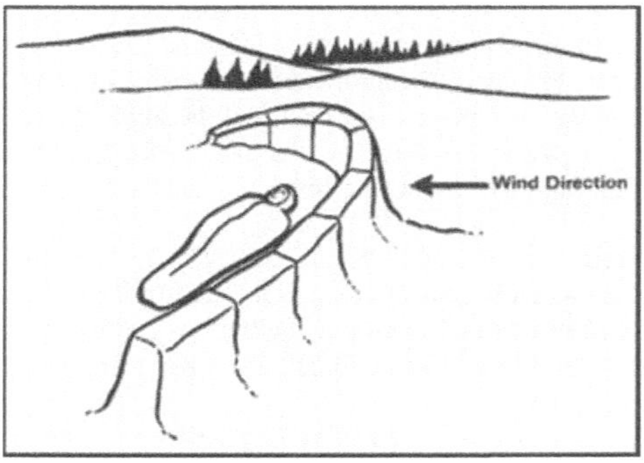

Figure 10-5. Snow wall.

SNOW TRENCH

A snow trench provides shelter quickly. It is built by burrowing into a snowdrift or by digging a trench in the snow and making a roof of ponchos or ice and snow blocks supported by skis, ski poles, or snowshoes. Boughs can be used for covering the roof and for the bed. Figure 10-6 shows one method of roofing over the trench by leaning snow blocks against each other.

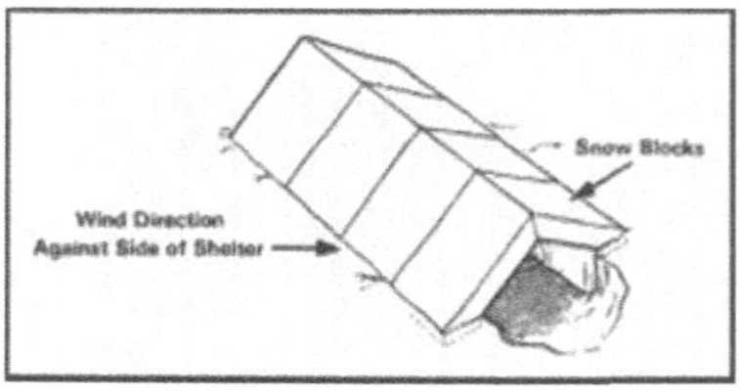

Figure 10-6. Snow trench.

SNOW CAVES

A snow cave can be used as an improvised shelter in the open areas where deep and compacted snow is available. Normally, a suitable site can be located on the lee side of a steep ridge or riverbank where drifted snow collects, and it is wind packed.

The tunnel entrance must give access to the lowest level of the chamber. This is the bottom of the pit, where cooking is done and equipment is stored. The snow cave must be high enough to provide comfortable sitting space. The sleeping area must be on a higher level than the highest point of the tunnel entrance, where there is warmer air.

The roof must be arched for strength so that drops of water forming on the inside will not fall on the floor. This water will follow along the curvature of the roof and sides, glazing over the walls when frozen. The roof must be at least 30 centimeters thick. The size of the snow cave depends upon the number of men that will occupy it. A large cave is usually warmer and more practical to build and maintain than several small caves.

The shape of the snow cave can be varied to suit conditions. When the main cave is built, short side tunnels are dug to make one- or two-man sleeping rooms, storage space, latrine, or kitchen space.

In building the snow cave, locate a deep snowdrift at least 3 meters deep. Newly fallen, powdery, or loose snow should not be used. The depth of a snowdrift may be tested with a sharpened branch or young tree about 4 meters long, or with the ski or shorter ski pole. The entrance must be chosen carefully so that wind will not blow into the cave or block the entrance by drifting snow. The entrance should be built so that it is about 45 degrees from the downwind side. A small tunnel is burrowed directly into the side of the snowdrift for 1 meter. A chamber is excavated from this tunnel. Excavation is done to the right and left, so that the length of the chamber is at right angles to the tunnel entrance. Personnel doing the digging will become wet from perspiration and from the snow inside the cave. They should wear the minimum amount of clothing to ensure that they have a change of dry clothing when finished.

The cave can be heated with the one-burner gasoline stove or with candles. The fires should be put out when personnel inside the cave are sleeping, thus reducing the danger of fire, asphyxiation, or carbon monoxide poisoning. If the weather is severe and it is necessary to have a fire while sleeping, an alert fire guard must be posted in each cave. The ventilation holes must be inspected every 2 or 3 hours to ensure that they have not become clogged by snow or ice.

To ensure that the cave is warm, the entrance should be blocked with a rucksack, poncho, or snow block when not in use. All available material such as ponchos, cardboard from ration boxes, brush, and boughs should be used for ground insulation.

Walking on the roof may cause it to collapse. At least two ventilator holes are used, one in the roof and one in the door. A ski pole can be stuck through the roof ventilator to clear it from the inside. Extra care must be taken to keep air in the cave fresh when cooking or heating. The entrance should be marked by placing a pair of skis, or other equipment, upright on each side of the entry way.

Building a snow cave requires practice. Large shelters are usually more efficient for squad-size units and larger units. The most critical decision is when and where to stop and build the shelter.

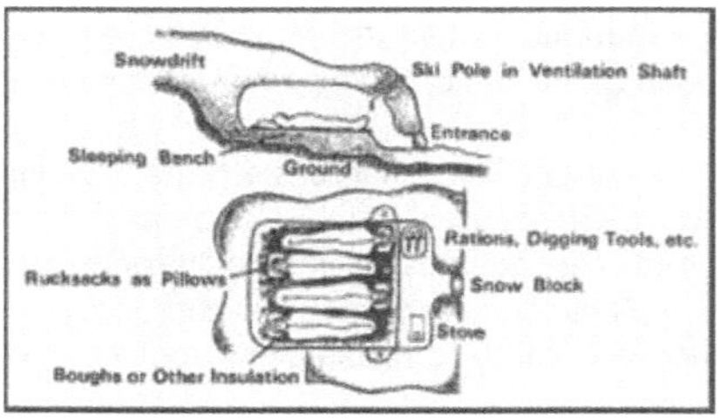

Figure 10-7. Snow cave.

INDIVIDUAL SNOW HOUSE

If you are alone and have need of a shelter quickly, you can build a small individual snow house. It is box-shaped, and the only building materials needed are snow blocks. Cut snow blocks 40 by 40 by 80 centimeters. Lay them on edge to a length of 2.5 meters and a width of 1 meter. Build the walls up to a height of about 1 meter. When you have the walls built, make an A-shaped roof by leaning together pairs of blocks. Fill in the ends with snow blocks and chink all the crevices with soft snow. Cut a hole in the end of the leeward side large enough to crawl into. When you are inside, make one or two small holes near the top for ventilation. Block up the end with a snow block or with your rucksack.

Figure 10-8. Individual snow house.

SNOW HOUSE OR IGLOO

An igloo makes an excellent shelter in snow and glacier areas when other materials are not available or when the snow is so hard that it can be cut into blocks. Cracks in the igloo are then covered with loose snow. To increase the head room and to prevent water droplets from forming inside the igloo, the corners of the blocks may be cut off to form a round and smooth interior.

A shelter may be built over the entrance in the same manner as the igloo, and it should be 1 to 2 meters long and about 1 meter wide and 1 meter high. This shelter provides more protection and warmth inside the igloo and a place to store equipment. Blocks of snow or other material may be placed at both the outside and the inside entrances of the tunnel for more warmth. Make a ventilation hole near the top and at the same height as the fifth or sixth row of snow blocks. The hole can be kept open by placing a ski pole or long stick into it and clearing it out as needed. The igloo is heated with a one-burner stove or with candles.

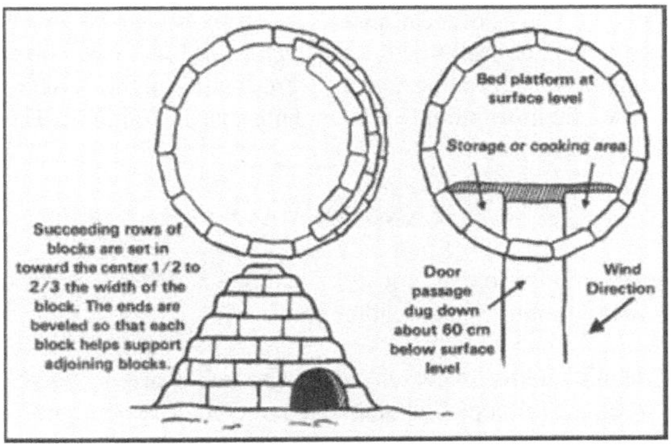

Figure 10-9. Snow house.

10-4. FOOD

The chances of finding different types of food vary in cold-weather areas depending on the time of year and the place. Shores normally are scraped clean of all animals and plants by winter ice. Food can still be found north of the timberline, even when such foods as mice, fish, and grubs are not available.

STORAGE AND PRESERVATION

If a large animal is killed or a large number of small game are found, you should store or preserve some of the meat for future use. Freezing is the best way to preserve fresh meat or fish. It may be necessary to protect your supplies from small animals. This can be done by hanging these supplies about 2 meters from the ground.

FISH

There are few poisonous fish in cold-weather areas, but some fish, like the sculpin, lay poisonous eggs. The black mussel may be poisonous at any season and is as deadly as strychnine. In cold regions do not eat shark meat; it is poisonous. In coastal streams and rivers, salmon moving upstream to spawn may be plentiful; however, their flesh deteriorates as they travel away from the coast, making them unfit to eat except in an emergency. A salmon that has been in fresh water long enough to turn a brilliant red is decaying and may be unfit to eat.

Coastal waters are rich in all seafood. Grayling, trout, white fish, burbot, and pike are common to the lakes, ponds, and arctic coastal plains of North America and Asia. River snails or freshwater periwinkles are plentiful in the rivers, streams, and lakes.

Fish can be speared, shot, netted, hooked, caught by hand, or stunned by a rock or club. Fish hooks can be improvised from pins, pieces of wire, or any other metal object that can be bent into a hook. Pieces of meat, insects, or minnows can be used for bait. Some fish will nibble at any small object that hits the water. Cod will swim up to investigate strips of cloth or bits of metal or bone, and they are easily caught. A net can be made out of stout twine or from the inner strands of parachute suspension lines. These strands can also serve as fishing lines.

LAND ANIMALS

Deer, caribou, reindeer, musk-oxen, moose, elks, mountain sheep, goats, bears, and other such animals are found in most cold-weather areas.

Tundra animals include rabbits, mice, lemmings, ground squirrels, wolves, and foxes. Where there are trees, the porcupine is often encountered. The porcupine can be easily clubbed or shaken out of trees. Pick up a porcupine only after it is dead.

Hunting is better during the early morning or late evening when the animals are moving to and from feeding and bedding grounds and water. Large game such as moose and caribou should be shot just back of the front shoulder or, if the range is short and you are sure of your shot, in the neck. These large animals are fairly easy to stalk and kill, and their skins are also useful for survival. Bears should be shot just forward of the front shoulder to stop them or, if you are sure of your shot, in the neck. To successfully hunt these land animals you should know something of their characteristics:

- Caribou or reindeer are usually very curious. It is possible to attract them near enough for a shot by waving a cloth and moving slowly toward them on hands and knees.
- The technique of imitating a four-legged animal may also cause a wolf to come closer to a hunter.
- Moose may be found in heavy brush or around the shoreline of lakes.
- Mountain goats and sheep are cautious and hard to approach. They can be surprised, however, by moving quietly downwind while they are feeding. If possible, stay on higher ground than they are on.

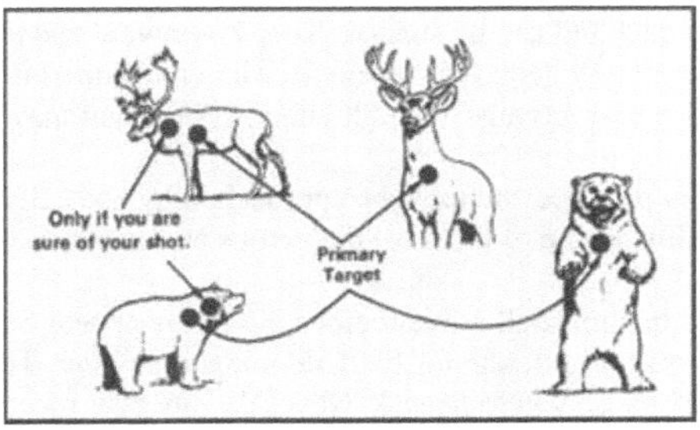

Figure 10-10. Hunting land animals.

- Musk oxen leave cattle-like tracks and droppings. When alarmed, they group together in a circle with their heads facing out and remain in that position unless approached, then one or more bulls may charge.
- Bears are always dangerous. A wounded bear is most dangerous and should not be followed into cover. The polar bear is a tireless, clever hunter with good sight and an extraordinary sense of smell. Be careful that you don't become the hunted instead of the hunter.
- Rabbits often run in circles and return to the same place when they are frightened. A running rabbit can sometimes be stopped by whistling. Snares made of wire and placed on small game trails, arranged so that the animal will place his head through the loop, are an excellent means of catching rabbits.
- The quick-kill method can be used on most small animals and sometimes on larger animals after they have been shot and are on the ground. The method is simple and consists of cutting the animal's throat with a sharp knife. This system not only kills the animal but bleeds him at the same time.

Figure 10-11. Snaring a rabbit.

SEA ANIMALS

During winter and spring, sea mammals — seals, walruses, and polar bears — are found on the frozen pack ice and on floes in open water.

Seals are hard to approach but can be stalked. Keep downwind and avoid sudden moves. A white camouflage suit will help. Advance only when the animal is sleeping. Do not eat the liver of the bearded seal, because its high vitamin A content may cause you to be sick.

Walruses are found on pack ice and must be approached by boat. Because of its large size and boldness, the walrus is one of the most dangerous animals in the arctic.

Polar bears are found in almost all arctic regions but rarely appear on land. Avoid them if possible. If it is necessary to kill one for food, do not eat the liver. The liver is high in vitamin A, which may cause you to be sick. Never eat any bear meat unless it is cooked; it is always diseased.

BIRDS

The breeding ground of many birds is in the arctic. Ducks and geese build their nests near ponds on the coastal plains during the summer and provide an important source of food. Grouse and ptarmigan live in mountainous terrain and brush-covered areas in arctic and subarctic areas. Sea birds may be found on cliffs or small islands on the coast. Sea birds, ravens, and owls are useful for food.

In winter, owls, ravens, grouse, and ptarmigans are the only birds available. Ptarmigans and grouse are easily approached and are very tame. Although they are hard to locate because of the protective coloring, they provide a good source of food because they can be killed with a stone or a club.

PLANTFOOD

Most plants that grow in cold regions can be eaten. The water hemlock and the baneberry are the only poisonous plants. The water hemlock is one of the world's most poisonous plants. It can be recognized by where it grows, always in wet ground, and by the following characteristics: a hollow, partitioned bulb at the base of a hollow stem, spindle-shaped roots, and a foul odor, which is very noticeable in the root and bulb. The water hemlock is abundant in marshes near southern beaches and around marshy lakes in interior river valleys. It is never found on hillsides or dry ground. The baneberry is normally found on hillsides or in forested bog areas. Some mushrooms are poisonous and should not be eaten. Although the cold-region areas have many kinds of berries that are edible, most will not be available in the winter.

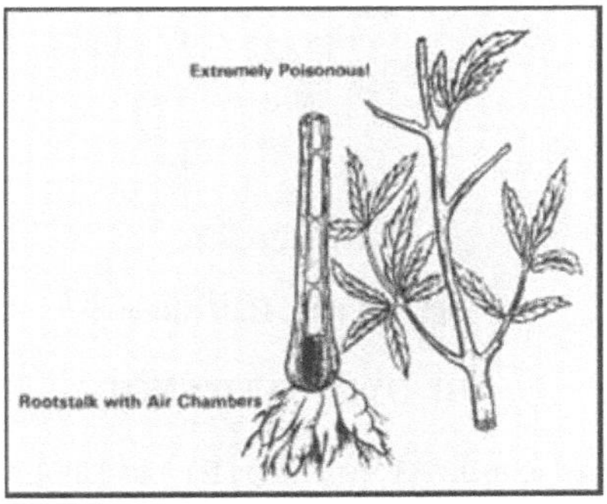

Figure 10-12. Water hemlock.

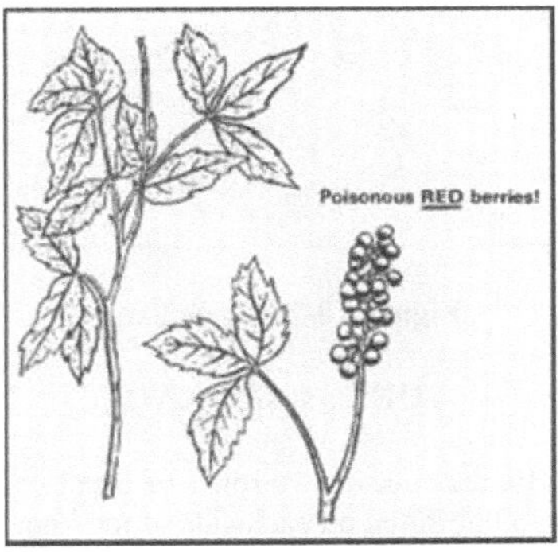

Figure 10-13. Baneberry.

10-5. USEFUL KNOTS

A rope is only a limp coil of hemp or nylon. It is useful only when you are able to attach it to itself or another object. The six basic knots described below will fit all your needs for tying rope together or onto equipment or personnel.

THE HALF HITCH

The half hitch is used to secure the end of a rope to some other object or to secure other knots.

Figure 10-14. Half hitch.

THE OVERHAND KNOT

The overhand knot is used to make a knotted rope for a handline or to temporarily whip the end of a rope to keep it from unraveling.

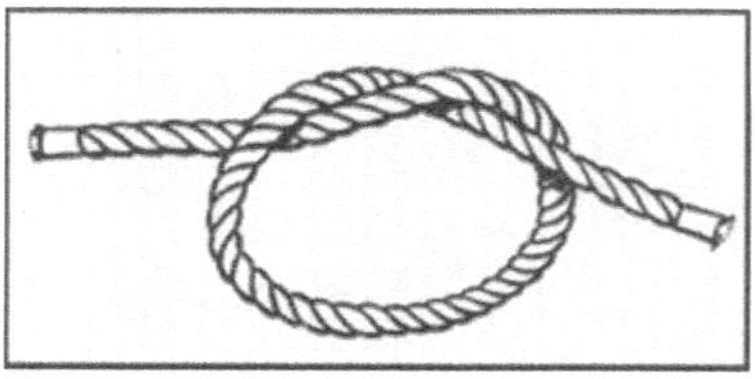

Figure 10-15. Overhand.

THE SQUARE KNOT

The square knot is used to tie the ends of two ropes of equal width together. The knot should be secured by tying a half hitch on each side of the knot. If properly tied, the square knot will not slip or jam. Be careful to follow Figure 10-16 or you may end up with a "granny knot," which can slip and jam.

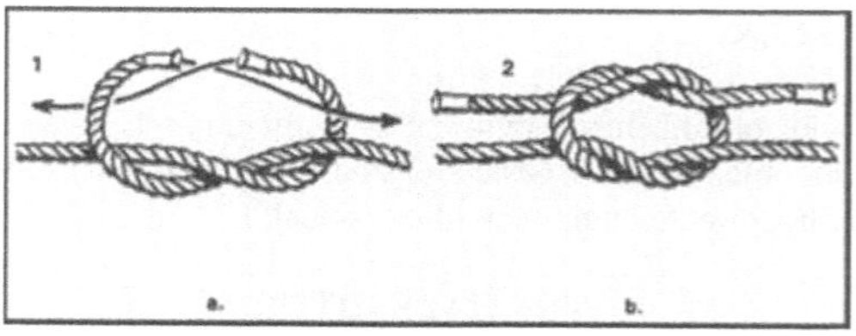

Figure 10-16. Square.

THE CLOVE HITCH

The clove hitch is the most useful of the hitches. It can be used to tie the end of a rope to a post or eye, or it can be used to secure the middle of a rope without using the ends.

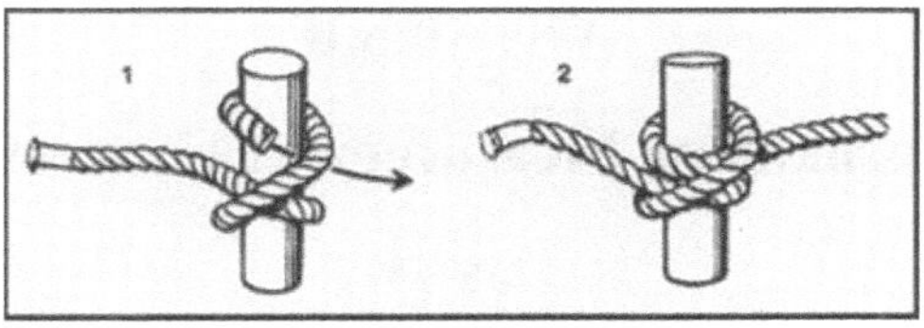

Figure 10-17. Clove hitch.

THE DOUBLE SHEET BEND

The double sheet bend is used for tying two ropes together of different width.

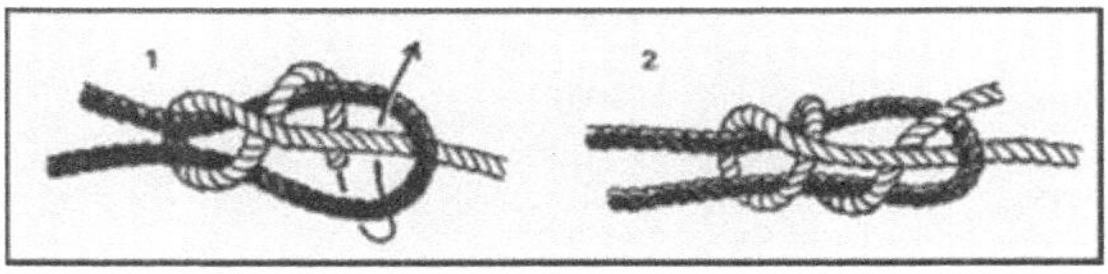

Figure 10-18. Double sheet bend.

THE BOWLINE

The bowline is used to form a loop on the end of the rope that will not slip.

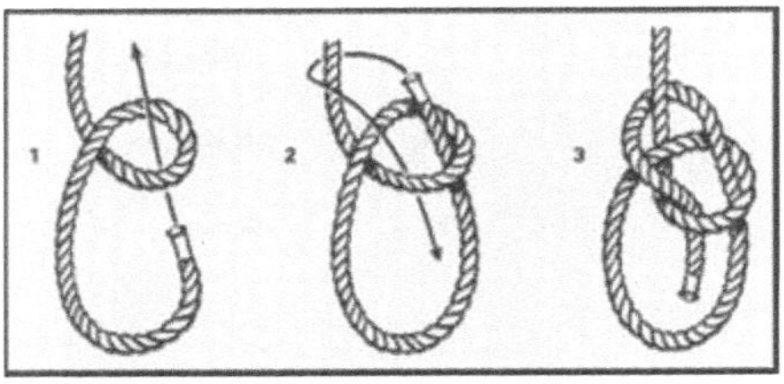

Figure 10-19. Bowline.

Remember the qualities of a good knot: it is easy to tie, it does not jam and become hard to untie, and it does not slip when weight is put on it.

APPENDIX B

Approximate Measurement Conversion

Factors

Multiply	*By*	*To Obtain*
Centimeters	.03	Feet
Centimeters	.4	Inches
Meters	3.3	Feet
Meters	40.0	Inches
Meters	1.1	Yards
Kilometers	.6	Miles
Knots	1.2	MPH
Inches	2.5	Centimeters
Feet	30.0	Centimeters
Yards	.9	Meters
Miles	1.6	Kilometers
MPH	.9	Knots

For ease in conversion the above factors have been rounded off.

References

Suggested Readings

DA Pam 20-201	Military Improvisations During the Russian Campaign
DA Pam 20-203	Russian Combat Methods in World War II
DA Pam 20-231	Combat in Russian Forests and Swamps
DA Pam 20-234	Operations of Encircled Forces: German Experiences in Russia
DA Pam 20-236	Night Combat
DA Pam 20-240	Rear Area Security in Russia: The Soviet Second Front Behind the German Lines
DA Pam 20-269	Small Unit Actions During the German Campaign in Russia
DA Pam 20-271	The German Northern Theater of Operations, 1940-1945
DA Pam 20-290	Terrain Factors in the Russian Campaign
DA Pam 20-292	Warfare in the Far North
FM 9-207	Operation and Maintenance of Ordnance Materiel in Cold Weather (0° to -65°F)
FM 31-70	Basic Cold Weather Manual
FM 31-71	Northern Operations
FM 90-6	Mountain Operations
FM 90-11 (TBP)	Winter Operations
TC 90-11-1	Military Skiing

"Perhaps the most valuable result of all education is the ability to make yourself do the things you have to do when it ought to be done, whether you like it or not."

Hosley

Glossary

ALICE	all-purpose lightweight individual carrying equipment
C	Celsius
CLP	cleaner lubricant preservative
ECWCS	extended cold-weather clothing system
F	Fahrenheit
FM	field manual
GP	general purpose
SFP	synthetic fiber pile
SOP	standing operating procedure
TBP	to be published
VB	vapor barrier

FIELD MANUAL
No. 7-85

HEADQUARTERS
DEPARTMENT OF THE ARMY
Washington, DC, 9 June 1987

FM 7-85
RANGER UNIT OPERATIONS

IMPORTANT

U.S. Army Infantry School Statement on
U.S. NATIONAL POLICY CONCERNING ANTIPERSONNEL LAND MINES

TABLE OF CONTENTS

Preface

CHAPTER 1. Introduction

1-1.	Mission	1-1
1-2.	Purpose and Function	1-2
1-3.	Fundamentals of Ranger Operations	1-2
1-4.	Capabilities and Limitations	1-4
1-5.	Employment Considerations	1-5
1-6.	Command and Control	1-6
1-7.	Communications	1-12

CHAPTER 2. Organization and Equipment

2-1.	Headquarters and Headquarters Company	2-1
2-2.	Regimental Headquarters	2-1

 DISTRIBUTION RESTRICTION: Approved for public release; distribution is unlimited.

2-3.	Regimental Headquarters Company	2-2
2-4.	Ranger Battalion	2-3
2-5.	Special Teams and Elements	2-5
CHAPTER 3.	**Deployment**	
3-1.	Readiness	3-1
3-2.	Ranger Ready Force	3-1
3-3.	N-Hour Sequence	3-2
3-4.	Intelligence and Security	3-2
3-5.	Operations	3-3
3-6.	Logistics	3-4
3-7.	Communications-Electronics	3-7
3-8.	Remote Marshalling Base and Intermediate Staging Base	3-7
CHAPTER 4.	**Insertion, Extraction, Escape, and Evasion**	
4-1.	Intelligence	4-1
4-2.	Deception	4-1
4-3.	Speed and Mobility	4-2
4-4.	Stealth	4-3
4-5.	Suppression	4-3
4-6.	Security	4-3
4-7.	Night Vision and Electronics Devices	4-3
4-8.	Rehearsals	4-4
4-9.	Sand Tables and Terrain Models	4-4
4-10.	Communications	4-4
4-11.	Planning Insertions	4-4
4-12.	Air Insertion	4-5
4-13.	Airborne Insertion	4-7
4-14.	Air-Land Insertion (Fixed-Wing)	4-12
4-15.	Air Assault Insertion	4-13
4-16.	Amphibious Insertion	4-16
4-17.	Land Infiltration	4-18
4-18.	Stay-Behind Techniques	4-21
4-19.	Extraction	4-21
4-20.	Escape and Evasion	4-22
CHAPTER 5.	**Strike Operations**	
5-1.	Application	5-1
5-2.	Raid Operations	5-1
5-3.	Interdiction Operations	5-7

5-4. Personnel and Equipment Recovery Operations 5-10

CHAPTER 6. Special Light Infantry Operations

6-1. Movement 6-1
6-2. Ambush Operations 6-4
6-3. Antiarmor Operations 6-14
6-4. Military Operations on Urbanized Terrain 6-14
6-5. Perimeter Defense 6-15
6-6. Breakout from Encirclement 6-17
6-7. Linkup Operations 6-18
6-8. Passage of Lines 6-19
6-9. Reconnaissance Platoon Operations 6-22
6-10. Reconnaissance Planning 6-23
6-11. Execution of Insertion 6-27

CHAPTER 7. Combat Support

7-1. Intelligence Support 7-1
7-2. Electronic Warfare Support 7-3
7-3. Communications Support 7-4
7-4. Fire Support 7-5
7-5. Aviation Support 7-14
7-6. Engineer Support 7-18

CHAPTER 8. Combat Service Support

8-1. Ranger Support Element 8-1
8-2. Logistics and Personnel 8-3
8-3. Resupply 8-8
8-4. Military Police 8-17
8-5. Counterintelligence 8-18
8-6. Personnel Service Support 8-18

APPENDIX A. Organization and Equipment of the Ranger Regiment A-1

APPENDIX B. Composition and Equipment of Ranger Regiment Liaison Teams B-1

APPENDIX C. Equipment Density C-1

C-1.	Communication Equipment	C-1
C-2.	Weapons and Equipment Each Battalion	C-2

APPENDIX D. **Standing Orders, Rogers' Rangers** D-1

APPENDIX E. **The Ranger Creed** E-1

APPENDIX F. **Ranger History** F-1

APPENDIX G. **Training** G-1

G-1.	Application and Guidance	G-1
G-2.	Philosophy	G-2
G-3.	Mission Analysis	G-4
G-4.	Unit Training	G-4
G-5.	Individual Training	G-4
G-6.	Collective Training	G-5
G-7.	Performance-Oriented Training	G-6
G-8.	Evaluation	G-6
G-9.	Field Training	G-7
G-10.	Professional Development	G-9
G-11.	Physical Training	G-9
G-12.	Environmental and Specialized Training	G-10
G-13.	Ranger Indoctrination Program	G-10

Glossary Glossary-1

References References-1

Authorization Authorization Letter

PREFACE

This publication contains doctrinal guidelines for the employment and training of ranger units. The intended audience is intermediate and higher level commanders, staffs of those headquarters, and newly assigned officers and NCOs of the ranger regiment. It is to be used with all other published US doctrine concerning NBC operations and the climatic or geographic influences on combat. The fundamentals apply to the employment of ranger units on worldwide operations and across the spectrum of conflict from low to high intensity warfare. This includes all aspects of the integrated battlefield: nuclear, biological, and chemical. It is not intended to apply to the employment of long-range reconnaissance patrol (LRRP) companies, even though such units carry the designation "ranger." These LRRP companies, as well as the long-range surveillance company (LRSC) (corps level) and the long-range surveillance detachment (LRSD) (division level), are organized, trained, and equipped for long-range reconnaissance in enemy territory.

The scope of this publication addresses the mission, organization, equipment, capabilities, limitations, planning, training, operations, and logistical support of ranger units. It describes how the ranger regiment and its subordinate battalions train, receive taskings, plan missions, task-organize, deploy, and integrate supporting elements to perform the mission. Once in combat, the ranger regiment maneuvers to accomplish the mission in the manner of other light infantry units. Ranger units make full use of published light infantry doctrinal guidance. This manual will repeat procedures or techniques contained in other publications only if necessary for understanding or continuity.

This publication does not contain everything the ranger needs to know to successfully execute a unit mission. Rather, it supplements unit training and the ranger unit commander's guidance; it does not supersede it.

The proponent of this publication is HQ, TRADOC. Submit changes for improving this publication on DA Form 2028 (Recommended Changes to Publications and Blank Forms) and forward it to Commandant, US Army Infantry School, ATTN: ATSH-B-ID, Fort Benning, GA 31905.

Unless otherwise stated, whenever the masculine gender is used, both men and women are included.

CHAPTER 1

Introduction

The ranger regiment is a major component of the US Army's special operations forces. It is a unique light infantry unit tasked to conduct special military operations in support of national policies and objectives. These operations require highly trained, well-disciplined units capable of employment in any environment, either alone or in concert with other military forces. Within this publication, the term *ranger force* describes any size force consisting mainly of members of the ranger regiment and led by a member of the ranger regiment's chain of command. A ranger force may be a TOE unit or it may be a specially organized task force for a specific mission.

1-1. Mission.

a. The mission of the ranger regiment is to plan and conduct special military operations. These operations are conducted by specially trained, equipped, and organized forces against strategic or tactical targets in pursuit of national military, political, economic, or psychological objectives. They may support conventional military operations or they may be performed independently when conventional forces cannot be used. (See Figure 1-1.)

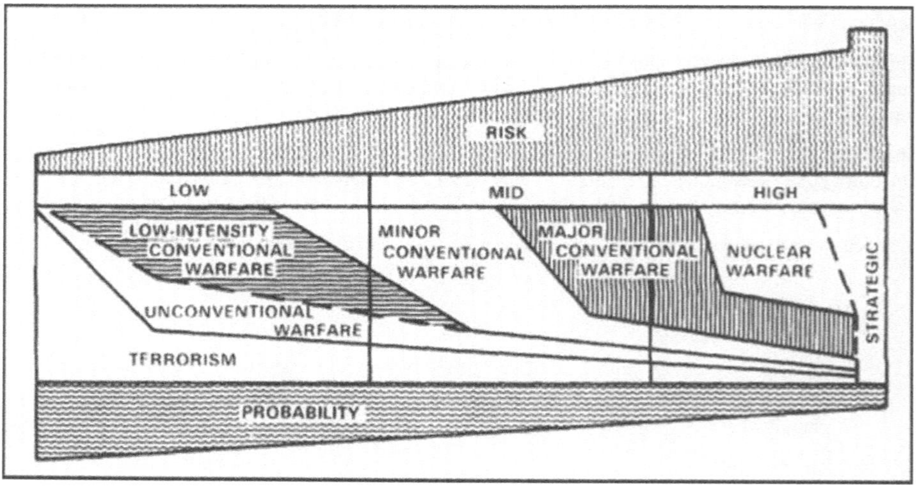

Figure 1-1. Spectrum of conflict.

b. Special military operations conducted by the ranger regiment include strike operations, usually deep penetration, and special light infantry operations. Strike operations include raids, interdiction, and recovery operations. Special light infantry operations include many of the light infantry missions assigned to airborne, air assault, or light infantry battalions and brigades. These operations are conducted in support of the AirLand Battle at all levels of intensity.

1-2. Purpose and function.

The ranger regiment provides the national command authority (NCA) the ability to move a credible military force quickly to any region in the world. The regiment uses the entire spectrum of intelligence support, from national systems to organic assets. Ranger units maintain a readiness posture that supports their immediate commitment to battle once deployed. They are often tailored for specific missions and may require augmentation from external sources. Tactical mobility may be augmented by USAF or Army special operations aviation (SOA) aircraft.

a. The ranger regiment conducts light infantry operations on the integrated battlefield as well as in low intensity conflicts. Special light infantry operations are conducted to accomplish certain contingency missions during peacetime or as part of the echelons above corps (EAC) battle.

b. The regiment conducts both strategic and operational missions. Its efforts are combined into an overall plan to destroy, delay, and disorganize the enemy, or to cause him to divert his attention and combat forces to rear area security. The limited number of ranger units and the diverse targets dictate a careful assignment of missions.

c. The mission, enemy, terrain, troops and time available (METT-T) factors govern the command and control arrangement under which the regiment operates. The ranger regiment or separate ranger battalions are normally assigned to the headquarters whose area of responsibility includes the regiment's operational area(s). The strategic or operational value of potential targets means that ranger units are normally employed at no lower than corps level. Employment at EAC is routine. The EAC headquarters employing ranger units could be the commander in chief's (CINC's) special operations command or field army. The controlling headquarters could also be that of a joint task force (JTF) or the Army component command (ARFOR) of a JTF. The ranger regiment's organization, communications equipment density, and training programs do not support operational level reconnaissance missions.

d. Ranger battalions are not oriented to a specific theater. Current force structure, contingency plans, and training needs (see Appendix G) preclude committing battalions to one region.

1-3. Fundamentals of ranger operations.

a. The success of an operation by a ranger unit depends on the observance of the fundamentals of the US Army's AirLand Battle doctrine (see FM 100-5). The ranger regiment fights the enemy as a light infantry force. It follows infantry doctrine and observes all the basic rules of infantry operations.

b. Along with initiative, depth, agility, and synchronization, ranger operations require--

(1) Detailed planning and coordination that allow the ranger unit to discern and exploit the enemy's weaknesses while avoiding his strength. Both permit the evaluation and use of information gathered from all sources. Detailed coordination integrates all supporting units and services. During planning, the ranger unit conducts pre-mission training, briefings, and rehearsals for all personnel. The use of special mission equipment is practiced and perfected.

(2) Decentralized execution IAW mission orders and the commander's intent. Special operations forces must use individual and unit initiative. Mission-type orders give the ranger force commander the flexibility to take advantage of opportunities on the battlefield.

(3) Surprise, achieved through the ranger unit's ability to move by uncommon means, along unexpected routes, and over rough terrain. Ranger units normally conduct operations during poor weather and reduced visibility, aided by night vision devices. This adds to the attainment of surprise.

(4) Survivability, achieved by using the classic infantry combat techniques of stealth and concealment. The ranger unit engages the enemy at the time and place of its own choosing. It takes full advantage of terrain, and destroys or suppresses enemy weapons. The ranger unit seeks to destroy or neutralize the enemy's command and control systems, his surveillance assets, and his mobility assets. Survivability is enhanced by rapid mission accomplishment and a prompt departure from the objective area.

(5) Mobility, speed, and violence of execution that allow the ranger unit it to close quickly on the objective area and complete the mission before the enemy can react. The speed at which events take place confuses and deceives the enemy as to the intent of the ranger unit. This forces the enemy to react rather than to take the initiative. Tactical mobility lets the ranger unit break contact and withdraw from the objective area.

(6) Shock effect, which is a psychological advantage achieved by the combining of speed and violence with the precision of the ranger attack. The ranger unit strives to apply its full combat power at the decisive time and place, and at the point of the greatest enemy weakness. The ranger unit strives to achieve maximum physical and psychological effect on the enemy by exhibiting aggressiveness and reasoned audacity.

(7) Multiple methods of insertion and attack, trying not to repeat operations thus decreasing the chance the enemy will detect a pattern. This is achieved through imaginative training and planning.

(8) Deception, practiced by the ranger unit during all phases of its operations, from deployment through the insertion phase, to the actions in the objective area and extraction. The enemy is kept unaware of the ranger unit's intentions, is fed conflicting and wrong information, and is kept from knowing the size or mission of the ranger force. The ranger force commander makes full use of ruses, feints, false insertions, electronic countermeasures, and dummy transmissions. Concurrent military operations by US or allied forces may be used as cover for ranger operations. Ranger units are careful to hide every possible aspect of an operation and to disguise those that cannot be hidden. While in the operational area, or when departing any location, ranger units ensure that no material is left that could provide the enemy with information concerning the ranger force.

(9) Audacity, achieved by a willingness to accept risk. The ranger force commander considers what the enemy expects the unit to do and the actual ranger abilities, He then chooses a course of action that may confuse the enemy, while remaining within the capability of the ranger unit. This fundamental is combined with deception and surprise to disrupt the enemy command and control.

c. Detailed planning and coordination finds the enemy. Surprise, mobility and speed, variety, deception, and audacity combine to shock and disorient the enemy, fixing him in place. The violence and precision of the ranger attack finishes the enemy while ensuring the ranger force survives.

1-4. Capabilities and limitations.

a. The ranger regiment has the following capabilities:

(1) Deploying quickly to conduct operations on all types of terrain and in all kinds of weather.

(2) Establishing a credible American presence in any part of the world to show US interest or resolve.

(3) Infiltrating and exfiltrating an area of operations, and assaulting an objective by land, sea, or air.

(4) Conducting strike operations to include raids, personnel and equipment recovery operations, and interdiction of key areas.

(5) Conducting special light infantry operations to include seizing and securing airfields, communications centers, command and control facilities, and key bridges; and other special light infantry operations.

(6) Performing short-duration reconnaissance of assigned ranger objectives for the ranger force commander.

(7) Operating for up to three days without resupply, and for longer periods when provided with accompanying or airdropped supplies.

(8) Providing liaison, communication, and coordination personnel and equipment to integrate the deployed ranger force into the logistical, intelligence, and operational system of the theater or joint task force (JTF) commander.

(9) Assuming operational control, for a limited time, of other US military forces such as engineers or infantry, airborne, or air assault battalions.

(10) Providing the focal point for all-source intelligence support to attached and assigned units of the regimental task force.

(11) Conducting limited combat operations under conditions of chemical, nuclear, or biological contamination.

b. The ranger regiment has the following limitations:

(1) Limited capability against armored or motorized units in open terrain.

(2) No organic transportation.

(3) Limited sustained combat capability due to the shortage of organic combat support and combat service support elements.

(4) Limited organic air defense weapons.

(5) Limited organic indirect fire support.

(6) No casualty evacuation capability.

(7) Reconstitution and retraining needed to replace combat losses.

1-5. Employment considerations.

a. Ranger units are characterized by the quality, motivation, training, and individual skill of their members. This produces units with superb collective abilities, able to adapt well to changing, complex situations.

b. Ranger units can conduct either deliberate or quick-response operations.

(1) Deliberate operations rely on careful planning, reconnaissance and surveillance of the target area, deception, secrecy, thorough preparation and rehearsals, and violent execution. A deliberate operation aims to complete the mission even though the enemy may have heavy forces on or near the objective area. Deliberate operations allow for detailed planning, evaluation, rehearsal, and coordination before insertion. A deliberate operation is likely to succeed against targets that the enemy has protected in depth, that have strong natural defenses, or that need a detailed and long insertion process.

(2) Quick-response operations rely on the high level of training and readiness of the ranger regiment to execute a mission before the enemy can react. These operations are conducted when there is little time for long, detailed planning. They rely on set procedures set forth by the ranger regiment and its supporting elements. A quick-response operation aims to complete the mission before an enemy can react. This type of operation may be chosen due to the time-sensitive nature of the target, political or military goals, the time frame of other operations, or the increased chance of enemy detection.

(3) Whether an operation is to be deliberate or quick response is often a difficult and time-sensitive decision by a high-level command authority. Decision-makers must consider the enemy's strength in the area, his intentions, his ability either to reinforce or to alter the target area, and the consequences of success or failure of diplomatic or military initiatives in related areas. The ranger unit commander tries to use the existing time, manpower, and resources to complete a detailed and coordinated plan. He refines that plan up to the insertion into the objective area.

c. Ranger units train to operate in any environment or weather condition. They regularly perform operations during periods of limited visibility. Ranger units maintain a high state of physical fitness and often train in close quarters combatives. In addition to completing advanced marksmanship training with standard US weapons, each member of a ranger unit trains with many foreign weapons. Ranger units are trained to operate on urbanized terrain, becoming specialists in entry and clearing techniques and quick-fire methods, especially during periods of limited visibility.

d. Because ranger units have limited vehicles, logistics operations capability, indirect fire support, and heavy weapons systems, they are not designed for continuous operations. During all phases of operations and training, ranger units need responsive external support.

e. Ranger units are normally employed against targets and under conditions that need their unique skills. Although targeting priority is set by the overall commander, ranger units are not normally assigned missions that can be done by conventional aerial bombardment or by other units.

f. Ranger units are oriented toward offensive operations. They are not normally employed as a rear area protection force. Although the ranger regiment has a small reconnaissance unit, ranger units do not normally conduct long-range reconnaissance missions. The structure, communications, and training of the ranger unit do not prepare it for LRRP missions. Ranger units engaged in strike and special light infantry operations have a secondary mission to collect and report combat information.

g. Ranger units can be deployed worldwide when US military presence or participation with a host national military activity would serve US interests. This deployment shows a readiness to commit forces into a threatened area or proves US national resolve. After the deployment, other activities include staging operations, rehearsing combat operations, securing base areas for use and deployment of other forces, and so on that provide a clear signal of US intent. Ranger units are not trained or organized to provide mobile training teams (MTT) to train indigenous forces. The US Special Forces or other special operation forces are trained to conduct such security assistance operations.

h. Ranger units can serve as an example to a host country and provide limited military advice and training. They will normally be augmented with linguists and technicians to increase their abilities. The regiment would normally still function as a unit. Most of its assistance would be through short-term, high-impact, unit-oriented operations. They would not be long-term individual efforts associated with advisory-type activities.

i. Ranger units may be deployed to engage in combined training exercises with allies. This enhances US national image by demonstrating the outstanding abilities of the American ranger. These activities may include ranger, light infantry, airborne, air assault, or amphibious operations.

j. Ranger units use standard US nuclear, biological, chemical (NBC) warning; detection; protection; and decontamination equipment and doctrine when operating on the integrated battlefield.

1-6. Command and control.

a. **External.**

(1) The ranger regiment is a key component of the US Army's special operations forces. The other elements of special operations forces are special forces (SF), psychological operations (PSYOP), civil affairs (CA), and special operations aviation.

(2) The 1st Special Operations Command (1st SOCOM) is responsible for command of all active component special operations forces elements, to include the ranger regiment, in peacetime. The 1st SOCOM has command and control of all assigned special operations forces, less operational command (OPCOM) of special operations forces units forward deployed. (See Figure 1-2.) The ranger regiment, while based in the continental United States (CONUS), is under the control of 1st SOCOM. During peacetime, the OPCOM of deployed ranger units is through channels chosen by the NCA through the Joint Chiefs of Staff (JCS) to the unified command.

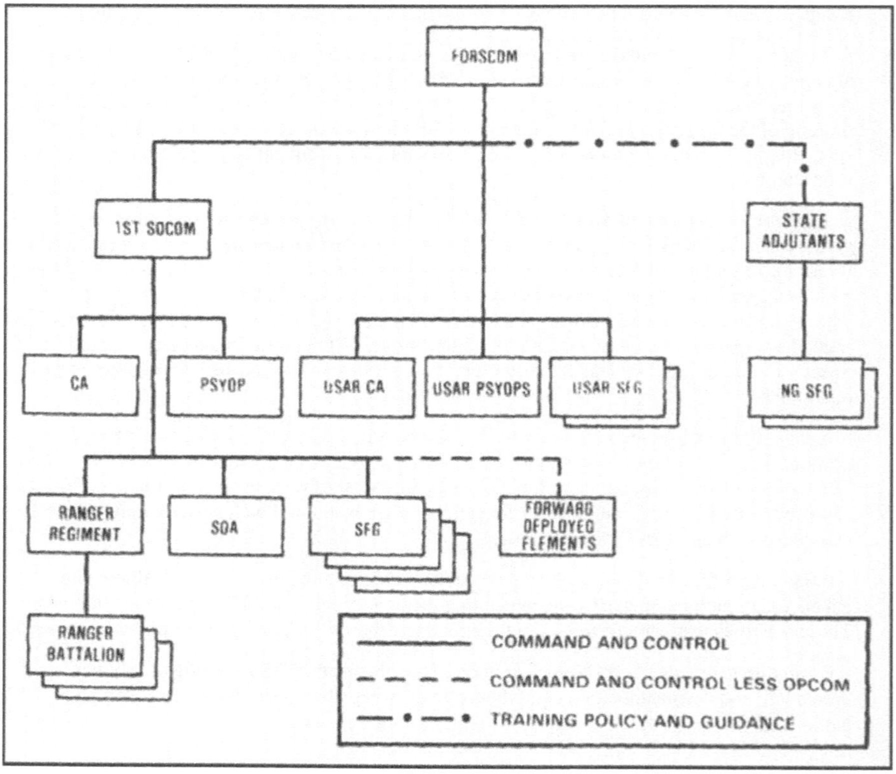

Figure 1-2. Command and control of Army Special operations forces during peacetime.

(3) In support of national policy, the NCA may direct contingency operations involving US

forces. These are politically sensitive military operations normally characterized by the short-term rapid projection or employment of forces in conditions short of conventional war. One aim of the NCA is to conclude them without going to war. The flexibility and effectiveness of the ranger regiment make it possible to use military force quickly to complete the mission and limit the spread of conflict. Command and control of special operations forces in a contingency operation is usually by a special operations task force (SOTF). (See Figure 1-3.) The SOTF and designated special operations forces are under the OPCOM of the highest level of command responsible for the operation. This command may be the unified command responsible for the contingency area (Option A), or a JTF designated by the NCA through the JCS (Option B). If no SOTF is formed, the ranger regiment could be OPCOM to the Army component commander (Option C). The ranger regiment or its battalions are not normally placed OPCOM to a division. A division's area of influence rarely contains appropriate targets. The division also lacks the assets to adequately support ranger operations. Employment of a ranger unit by a corps-level command is normally on a case-by-case basis. The theater or JTF commander retains centralized control over the mission assigned to the ranger unit.

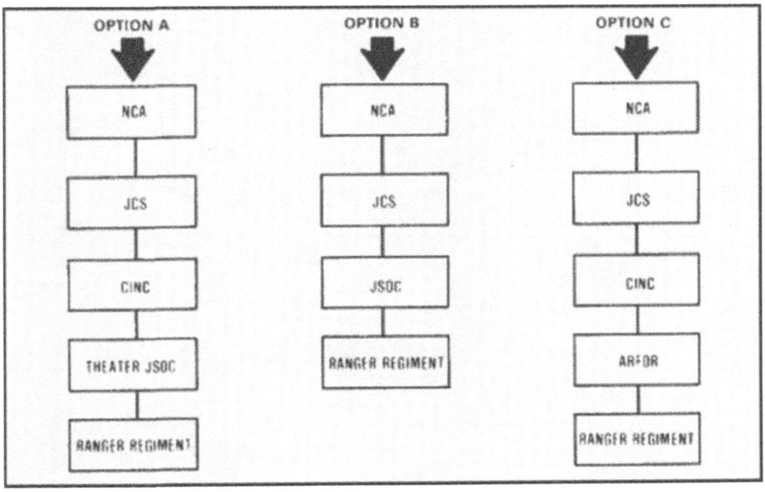

Figure 1-3. Command and control of Army Special operations forces during contingency operations.

(4) The 1st SOCOM is responsible for the deployment of the ranger force to the theater area. The headquarters having OPCOM provides the all-source intelligence, secure communications, insertion and extraction assets, and logistical support needed. The responsible commander in chief, through the Army component commander, provides support during the contingency operation.

(5) During wartime, ranger units are deployed as strategic assets to the theater of operations.

Strategic mission support is provided to the unified commander (CINC) from the NCA through the Joint Chiefs of Staff. It may involve the execution of sensitive operations. The employment of ranger units is controlled at the highest level of command directly responsible for the mission. Ranger units are assigned to a REMAB based on mission, force protection, and OPSEC considerations. The employing command and other elements provide timely and accurate intelligence, communications, logistical, and administrative support. Command and control of these units conforms to the provisions of JCS Pub 2. (See Figure 1-4.)

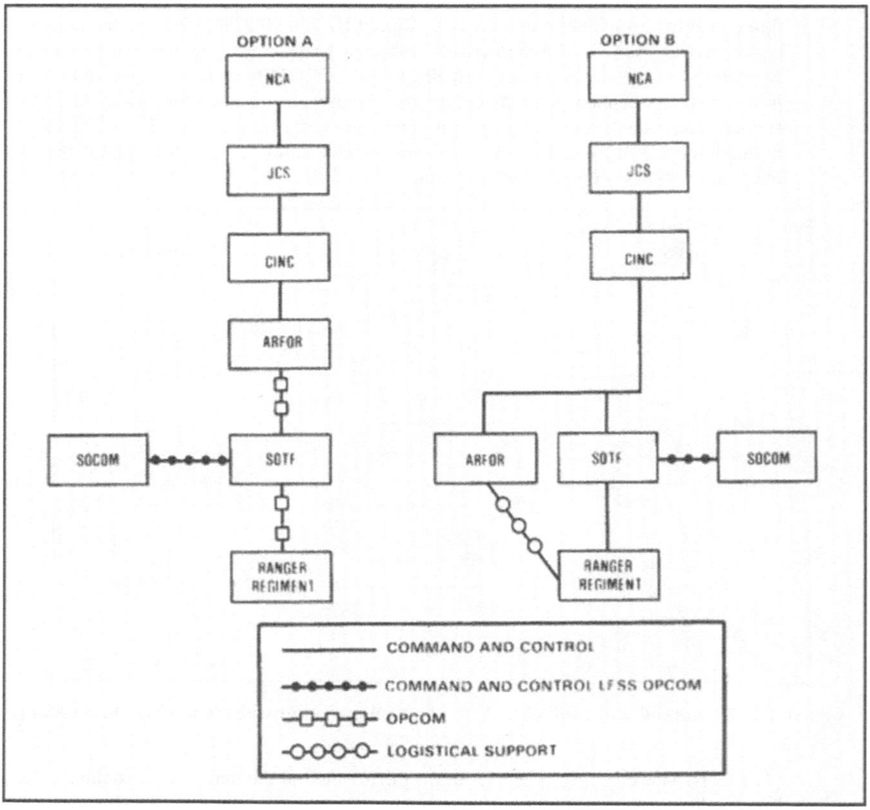

Figure 1-4. Command and control of Army Special operations forces during wartime.

(a) A theater joint special operations command (JSOC) is subordinate to the unified command to supervise the command and control, employment, and support of all US and allied special operations forces. The JSOC is a key element in the command and control of ranger units performing strike missions or special light infantry operations (Option A).

(b) The JSOC may be placed on a lateral line with other in-theater service components (Option B). This can occur when ranger units are -employed far from conventional forces and against strategic targets. It can also occur when the JCS or a CINC forms a JTF to conduct a limited-duration operation in an area where ranger units and conventional forces may operate near each other, or where operations must be coordinated on a regional level.

(c) Under the Army special operations command (ARSOC) concept, 1st SOCOM deploys as a command and control headquarters within the theater. When this occurs, the command relationships will be as shown in Figure 1-5.

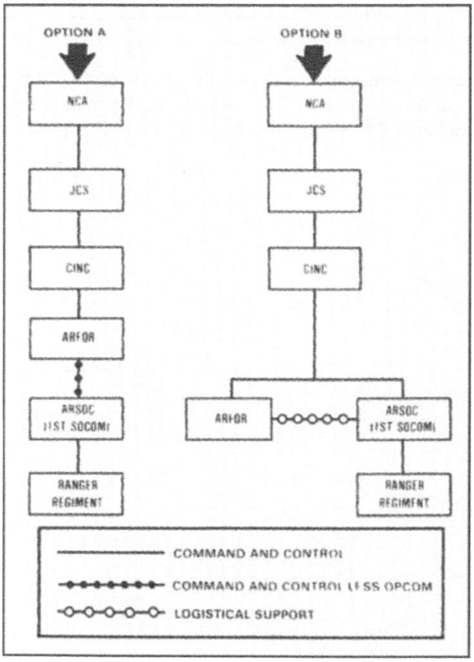

Figure 1-5. Command and control of Army Special operations forces under the ARSOC concept.

(d) Recently passed legislation will modify existing high level SOF command and control arrangements. All US Army, USN, and USAF special operations forces will be consolidated into joint commands at both national and theater level. The ranger regiment will continue to operate under the command of the 1st SOCOM, while increasing its ability to conduct joint operations.

(6) Ranger units conducting strike or special light infantry missions are not normally offered for transfer of authority (TOA) to allied forces. However, this will not preclude allied commands and US elements from combined planning and targeting for ranger missions.

(7) Requests for ranger units, from either US or allied commands, go through normal command channels to the CINC. The CINC, through the JSOC, is responsible for the employment of all special operations forces, including ranger units. This provides flexible response to JCS-directed missions, ensures quick response to the CINC's needs, enhances the joint employment of special operations forces, and lessens the chance of overcommitting and degrading the ranger force.

(8) Command and control of the ranger regiment and its battalions is IAW AirLand Battle doctrine. Before commitment to a mission, command and control is kept at a level where the unit's unique skills can be applied worldwide. Once committed, the ranger unit is placed under the operational control of the command that is responsible for the mission. This normally corresponds to the command whose area of influence includes the target area. Ranger task forces may operate directly under the authority of the senior American representative in the country. Command and control arrangements may vary according to the theater involved and METT-T considerations.

(9) The command and control of special operations is centralized and begins at the highest level of the military and civilian decision-making process. The specific command relationships are normally based on the mission.

b. **Internal.**

(1) The ranger regimental headquarters is organized to operate like that of a brigade headquarters. In addition to commanding and controlling all three ranger battalions, the regimental headquarters can assume operational control (OPCON) of conventional combat and combat support units, and other special operations forces for limited periods. It provides the regimental commander with a well-balanced responsive staff. It is structured as an operational headquarters that can deploy to the mission area and act as the ground tactical headquarters.

(a) The regimental headquarters prepares for combat by forming two small command and control groups. Each of these can deploy to an objective area and control combat operations. Other elements of the staff operate the tactical operations center (TOC) and the logistical operations center (LOC). These may be collocated depending on the situation. The TOC and LOC are normally located well behind the FEBA, at the REMAB or at the intermediate staging base (ISB). The ranger regiment also has two small liaison teams consisting of LNOS, staff representatives, and communications elements. (See Appendix B.)

(b) When a single ranger battalion is committed to an operation, the regimental commander normally exercises control from the TOC. A liaison and communications team is detached from the regimental headquarters to augment the ranger battalion. This team is attached to the deployed battalion until the end of the mission.

(c) If two or more ranger battalions are used in a single operation, the regimental commander normally deploys both command and control groups. He commands from the objective area, The use of two small, mobile control elements makes up for the possible loss of the regimental commander and the primary control element.

(d) Whenever the regimental commander is controlling an operation from the REMAB or the objective area, he normally provides a liaison team with a communications element to the higher headquarters exercising operational control. This cell can provide ranger representation to the controlling staff, operations interface, logistics interface, fire support planning and coordination, intelligence analysis and dissemination, and communications between the higher headquarters and the ranger battalion.

(2) The ranger regimental headquarters can serve as a ranger or special operations force task force headquarters. With augmentation, it can function as the Army special operations force command element. The regimental headquarters can also function as the Army component command of a JTF.

1-7. Communications.

a. Ranger missions need unique and dedicated communication support assets. Along with the ranger regiment's organic communications means, communications support assets are organized within the 1st SOCOM that-can deploy in support of the ranger regiment. This ensures that there are reliable communications to all elements of the SOTF. A special operations communications battalion (SOCB), deployed by the 1st SOCOM, provides more communications systems to support the special operations plan. The SOCB does not replace or duplicate existing in-theater communications. It augments the CINC's communications with the US Army's portion of the critical special operations force command, control, and communications link unique to the SOTF.

b. The ranger regiment's communication platoon can provide secure, long-range communication links from the CINC to a deployed ranger force. It can also enter the SOCB and the joint communications support element (JCSE) nets to provide a direct link with the NCA. Its equipment consists of four light vehicles equipped with various multiple communications systems. The two regimental liaison teams use one of these vehicles each. The TOC and LOC each use one.

c. Secure satellite communications (SATCOM) and radios with AM or FM capability are the main means of communication within the ranger regiment. These are also used for communications within the ranger battalions. The regimental commander has two light vehicles equipped with multiple communications systems. One of these is used to support each of the two mobile command and control elements of the regimental headquarters. Some of the radio systems can be dismounted and man-packed.

d. The ranger regimental headquarters has both man-packed and vehicular tactical satellite communication systems. Each of the ranger battalions has one base station and three man-packed systems.

(1) The man-packed radio is battery-operated. It provides two-way communications in both line-of-sight (mobile) and satellite (at-halt) modes. The digital message device group (DMDG) is the input and output device for data transmission. Secure voice communications is achieved using various encryption devices.

(2) The vehicular net control station uses the basic receiver/ transmitter unit installed on a vehicle. Each can serve up to 15 terminals in a communications net.

e. The reconnaissance platoon is equipped with multiple, man-packed communications systems. These let its teams operate in most of the communications nets within the regiment. Each five-man reconnaissance team has this communications ability, as does the platoon headquarters.

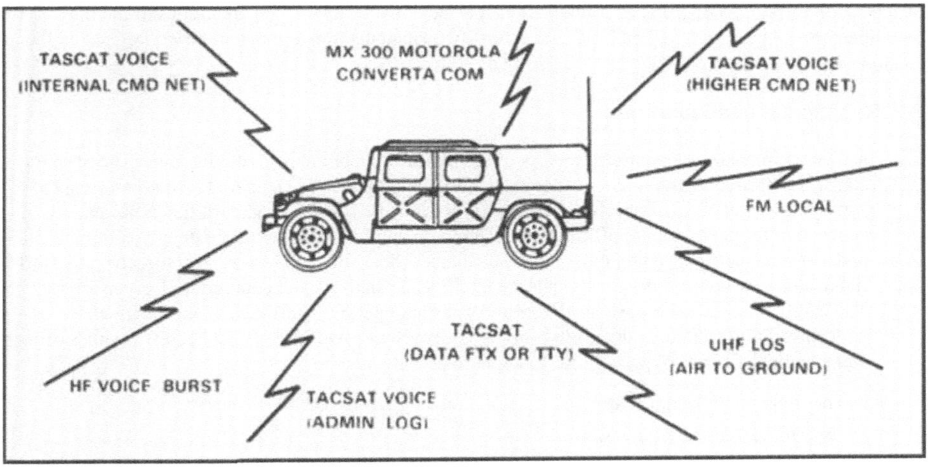

Figure 1-6. Communications capabilities within the regimental headquarters.

CHAPTER 2

Organization and Equipment

The ranger regiment is a specially organized and trained light infantry force consisting of about 2,000 personnel. It includes a regimental headquarters and headquarters company and three ranger battalions. (See Appendix A.) The regiment can operate as a single unit, separate battalions, or separate companies on specific missions. Independent missions are not normally assigned to ranger platoons.

2-1. Headquarters and headquarters company.

The HHC of the ranger regiment consists of 130 personnel, the majority being senior noncommissioned officers, commissioned officers, or soldiers with special skills to support the operations of the ranger battalions. (See Appendix A.) The HHC consists of two elements: the regimental headquarters and the headquarters company (minus).

2-2. Regimental headquarters.

a. The regimental headquarters consists of the regimental commander and his staff. The primary staff is composed of five sections: personnel (S1), intelligence (S2), operations and training (S3), supply (S4), and civil affairs (S5). These staff sections perform the standard functions and are supervised by the executive officer (XO). The regimental commander has a deputy commander and a special staff consisting of a communications officer, fire support officer, surgeon, and staff judge advocate. The S2 officer has a large intelligence analysis and counterintelligence section. The S3 section has extra liaison officers and assistant operations officers to coordinate the widespread and complex ranger missions. A United States Air Force (USAF) staff weather officer and a tactical air control party (TACP) are permanently attached.

b. The regimental headquarters:

- Analyzes assigned missions.
- Plans, coordinates, and controls training and operations.
- Coordinates the infiltration, resupply, exfiltration, and debriefing of the reconnaissance teams.
- Provides liaison support.
- Provides administrative, medical, NBC, and logistical planning and coordination support to the committed ranger battalions.
- Coordinates special support requirements (rigger, transportation, communications security [COMSEC]) for the ranger battalions.

2-3. Regimental headquarters company.

The regimental headquarters company (minus) consists of the company commander and his administrative personnel, plus the limited combat support (CS) and combat service support (CSS) elements assigned to the regiment. These are the fire support element, the communications platoon, the reconnaissance platoon, and the medical treatment team.

a. The company headquarters:

(1) Provides command and control of assigned elements.

(2) Provides administrative, personnel, logistical, and maintenance support to the assigned elements.

b. The fire support element:

(1) Plans fires to support the commander's ground tactical plan.

(2) Coordinates the integration of all available fire support means (mortars, artillery, armed helicopters, close air support, naval gunfire) to support mission execution.

c. The communications platoon (see Appendix A):

(1) Provides secure SATCOM teletype (TTY) and facsimile transmissions (FAX) abilities at the remote marshalling base or intermediate staging base (REMAB or ISB).

(2) Provides secure SATCOM, TTY, and FAX to the regimental headquarters and ranger battalions in the objective area.

(3) Provides cryptographic support to the regimental headquarters.

(4) Provides limited repair of radio and cryptographic equipment.

d. The reconnaissance platoon (see Appendix A):

(1) Trains, equips, and controls the employment of three reconnaissance teams.

(2) Infiltrates the objective area by parachute (high-altitude, low-opening [HALO]; high-altitude, high-opening [HAHO]; static line), helicopter, fixed-wing aircraft, self-contained underwater breathing apparatus (SCUBA), small boat, or other means.

(3) Remains undetected in the objective area for up to five days.

(4) Performs reconnaissance.

(5) Performs demolition target analysis.

(6) Uses small boats and inflatable rafts.

(7) Emplaces and monitors unattended ground sensors.

(8) Uses night image enhancement devices.

(9) Uses photographic equipment.

(10) Emplaces electronic target acquisition and designation devices.

(11) Collects specific combat information to satisfy priority intelligence requirements (PIR).

(12) Reports information using secure, long-range communication equipment.

Figure 2-1. Reconnaissance platoon communications.

(13) Performs drop zone (DZ) and landing zone (LZ) selection, marking, and reception duties.

(14) Escapes and evades enemy areas.

(15) Performs limited attacks or ambushes when tasked by the regimental commander.

(16) Reports weather in the objective area.

e. The medical treatment team:

(1) Provides medical support to the regimental headquarters.

(2) Coordinates required medical support and evacuation.

2-4. Ranger battalion.

a. The ranger battalion is organized and equipped to conduct both special and conventional combat operations. It contains organic combat, CS, and limited CSS elements. It consists of a battalion headquarters, a headquarters company, and three ranger rifle companies. (See Appendix A.)

b. The ranger battalion is the combat element of the ranger regiment. It can conduct independent combat operations or subdivide into its separate ranger rifle companies and their subordinate elements who can conduct semi-independent operations.

c. The ranger battalion headquarters and headquarters company (see Appendix A) includes:

(1) **Battalion headquarters.** The battalion headquarters consists of the commander; the executive officer; the command sergeant major; the S1, S2, S3, S4, and S5 officers; the chaplain; the communications officer; the fire support officer; and two radio operators.

(2) **Battalion headquarters section.** The battalion headquarters section consists of the S3 air and assistant S8, an assistant S2, an intelligence sergeant and a senior intelligence analyst, two operations sergeants and an operations assistant, a personnel staff noncommissioned officer (PSNCO), the chaplain's assistant, two clerk typists, two radio operators, a chemical officer and NCO, a personnel administration center (PAC) NCO, three PAC clerks, and a legal specialist. There is a USAF tactical air control party permanently attached.

(3) **Communications section.** The communications section consists of one telecommunications technician, one tactical communications chief, two combat signal radio team chiefs, one tactical communications systems supervisor, three radio operators, one switchboard operator, one single channel radio operator, and two equipment mechanics. The section provides communications support and associated maintenance.

(4) **Battalion medical section.** The battalion medical section consists of one general medical officer, one physician's assistant, two medical specialists, and twelve aidmen. The section provides unit-level medical support and trains selected members of the battalion in advanced medical procedures.

(5) **Battalion support platoon.** The battalion support platoon consists of a battalion support platoon leader, an ammunition NCO, and a food service section. The platoon provides limited CSS for assigned and attached units. This platoon does not have organic transportation. The battalion food service section consists of a food service sergeant (E7), a first cook (E6), and eight cooks.

(6) **Fire support team headquarters.** The fire support team (FIST) headquarters is supervised by the battalion fire support officer. It consists of the fire support sergeant and two fire support specialists. The headquarters controls three 9-man fire support teams. These teams are normally with the ranger rifle companies in a habitual support relationship. When operations are conducted near water, the fire support team headquarters may be augmented by spotter teams from the naval shore fire control party.

(7) **Company headquarters.** The HHC company headquarters consists of a commanding officer, an executive officer, a first sergeant, a supply sergeant, and an armorer.

d. The ranger rifle company (see Appendix A) includes a company headquarters, three rifle platoons, and a weapons platoon. Each company can conduct semi-independent operations for a short time when augmented by assets from the ranger battalion headquarters company.

(1) **Ranger rifle company headquarters.** The ranger rifle company headquarters consists of a commanding officer, an executive officer, a first sergeant, a supply sergeant, a tactical communications chief, two radiotelephone operators, an armorer, and an NBC NCO. It also has a two-man sniper team equipped with a sniper weapon. It normally has one of the 9-man FISTs attached from the ranger battalion headquarters company. Elements of this FIST may be further attached to the ranger rifle platoons.

(2) **Ranger rifle platoon.** The ranger rifle platoon consists of a headquarters section, three rifle squads, and a machine gun squad. The headquarters section personnel include a platoon leader, a platoon sergeant, and a radiotelephone operator. Each rifle squad consists of a squad leader and two balanced fire teams of four men each. The machine gun squad consists of a squad leader and three M60 machine gun crews of three men each. The machine gun squad is normally employed as a single element to provide quick, accurate, long-range supporting fire. One or more crews may be attached to other squads within the platoon.

(3) **Ranger weapons platoon.** The ranger weapons platoon consists of a headquarters section, a mortar section, and an antiarmor section. The headquarters section consists of a platoon leader, a platoon sergeant, and a radiotelephone operator. The mortar section has two 3-man 60-mm mortar squads. The mortar squads are normally employed as a section. They may be attached by section or squad to other platoons. The antitank section normally consists of a section sergeant and three 3-man 90-mm recoilless rifle or Dragon missile teams. They are normally employed as a section. The antiarmor section may also be attached by section or team to other platoons.

2-5. Special teams and elements.

The ranger regiment forms special teams to do critical tasks. These teams are made up of members of the unit and are used when needed. Examples of these special teams and elements are--

a. SCUBA teams whose members are graduates of the Special Forces Underwater Operations School. They conduct special insertion and reconnaissance duties.

b. HALO teams whose members are graduates of the Special Forces Freefall School and who conduct high-altitude, low-opening parachute operations.

c. HAHO teams whose members are also graduates of the Special Forces Freefall School and who conduct high-altitude, high-opening parachute operations, allowing offset parachute insertions.

d. Scout swimmer teams whose members are specially trained by the Landing Force Training Command (United States Marine Corps [USMC]) to conduct operations along coastlines, waterways, and inland waters. These teams conduct special insertion, reconnaissance, and security tasks often in support of amphibious operations.

e. Sniper teams, whose members are authorized by the table of organization and equipment (TOE), are used for both conventional and special operations. Along with the two-man sniper teams in the rifle company headquarters, each rifle platoon has two trained snipers.

f Demolition teams whose members are trained in advanced demolition and incendiary techniques and the use of special explosives,

g. Air defense teams trained in the use of man-portable air defense system (MANPADS) weapons. Such teams give the ranger battalion limited self-defense against air attack.

h. Liaison teams that are used by the ranger regimental commander to augment the existing command and control links from the employing unit headquarters to the ranger regiment. These teams are placed where they are most needed and may assist the higher headquarters staff in planning and target selection, as well as performing other liaison duties. See Appendix B for a detailed listing of the personnel and equipment included in the ranger liaison teams.

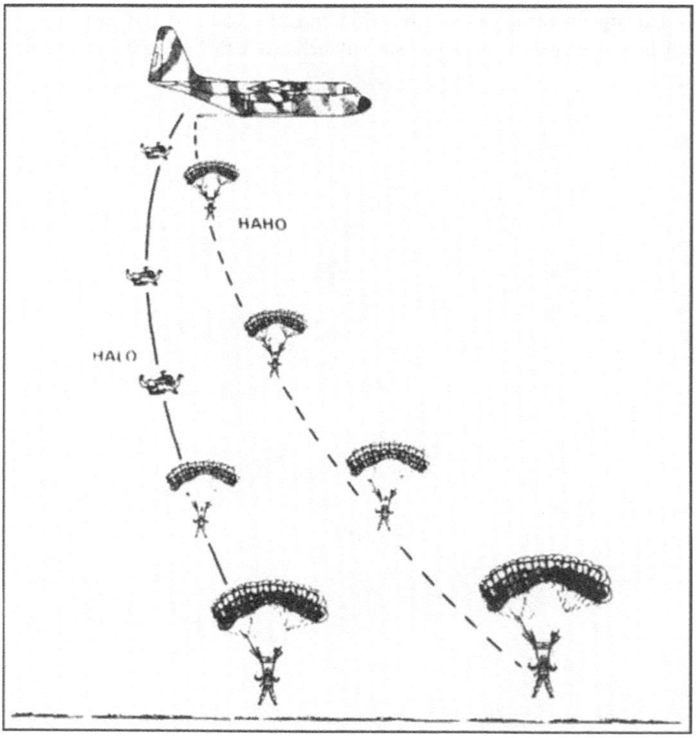

Figure 2-2. HALO and HALO parachute operations.

i. The ranger support element (RSE).

(1) The RSE is a special support element provided by host installation commanders at CONUS posts. It consists of elements and teams from either TOE units stationed at the installation or from the table of distribution and allowances (TDA) organizations under the command of the installation commander.

(2) The RSE provides the necessary support to ensure a timely and efficient deployment of the ranger battalion. During CONUS deployments, the RSE is responsible for both the departure and the reception support provided the battalion. During OCONUS deployments, the RSE is responsible for initial deployment and continuing support until the deploying ranger force is OPCON to the supported unified command or JTF. The responsible theater commander assumes support of the ranger battalion at the REMAB or the ISB if outside continental United States (OCONUS). On deployments that require the ranger battalion to move directly to the objective area from CONUS, the RSE continues support of the ranger battalion until released, and is prepared to accept considerable augmentation assets. The RSE provides the following support:

(a) Furnishing meals that are either served by a supporting unit in an established facility or taken to the ranger battalion at an isolated location.

(b) Issuing supplies from existing stocks, or from contingency items stored for use by the ranger battalion.

(c) Transporting supplies, food, personnel, and equipment.

(d) Palletizing equipment and ammunition for air movement and airdrop.

(e) Rigging vehicles and equipment for airdrop.

(f) Refueling vehicles and equipment.

(g) Issuing barrier and construction material for building of rehearsal sites.

(h) Helping in the construction of rehearsal sites and targets.

(i) Securing, storing, moving, and issuing ammunition and other Class V items.

(j) Maintaining ranger weapons and equipment, to include communications equipment. This is normally done by maintenance contact teams.

(k) Setting up and operating the RSE emergency operations center.

(l) Securing the REMAB if it is in CONUS.

CHAPTER 3

Deployment

Ranger units can quickly deploy by any means of transportation to any location in the world.

3-1. Readiness.

One ranger battalion is always in an advanced readiness condition as the Ranger Ready Force (RRF) available for immediate worldwide deployment. A second battalion is prepared to deploy later and a third battalion to follow. The regimental headquarters maintains command and control, liaison, communications, and reconnaissance elements immediately available for deployment. Higher status of readiness in response to specific world situations can be achieved.

3-2. Ranger ready force.

In order to maintain advanced readiness conditions, ranger battalions are rotated as the RRF.

a. Before taking over the duties of the RRF, battalions must do the following:

(1) Complete preparation for oversea movement (POM) qualification, to include an individual records check and required inoculations.

(2) Test fire all weapons, confirm battlesight settings, and repair or replace faulty weapons.

(3) Assemble and load on pallets the unit basic load (Classes I and V) and other supplies and equipment.

(4) Check unit-derived packing list components for accountability and serviceability.

(5) Prepare special operations troop listings.

b. During an RRF period, ranger units:

(1) Have a positive recall system for all personnel.

(2) Store enough air items for a battalion-size parachute assault.

(3) Conduct no off-post exercises, unless approved by the regimental commander.

c. Each unit keeps an alert roster and a telephone notification system for recalling personnel. The rosters ensure that the duty officer and the unit charge of quarters make a minimum number of calls. Each commander briefs personnel assigned to his unit on the recall system and tests it often.

d. Ranger units can maintain an increased readiness condition for long periods, but they should not be kept in that condition longer than needed. The effort to maintain this level of readiness will degrade the unit's training and effectiveness. If the needs of the operations dictate long periods at the highest readiness condition, provisions must be made for sustainment training. Also, added support will be required from the RSE.

e. The ranger regiment maintains a strong family support system when the unit is training or deployed on combat operations. Each member of the chain of command is responsible for family support.

3-3. N-hour sequence.

The N-hour sequence is a departure time sequence designed to help the ranger force complete all required deployment actions within a certain time. The length of the N-hour sequence depends on the readiness condition of the battalion before deployment. The N-hour sequence begins when the battalion is alerted and ends when the first elements depart. The N-hour sequence does not include travel time to the objective area.

3-4. Intelligence and security.

a. **Essential elements of friendly information.** Items of sensitive information and subject areas, if not secured, can be collectively analyzed by a hostile force. This analysis can reveal the intentions of a unit or compromise a mission. These items of information are designated as essential elements of friendly information (EEFI). They are never transmitted without encryption. The EEFI associated with a ranger unit alert normally are the mission, the use of staging bases, nature of the target, special terrain or weather conditions, and time limits. The ranger force commander must ensure that EEFI are not discussed outside of secure areas or made available to personnel without a security clearance and a need to know.

b. **Security plans.** Security plans for billets, marshalling areas, and out-loading areas must remain flexible to adapt to any situation. Certain items are always addressed in security plans. These are access control and physical security.

(1) **Access control.** Ranger units control access to their respective areas by allowing only authorized personnel to enter or leave. All entrances are secured and an armed guard with the access roster is posted. Access to telephones is restricted.

(2) **Physical security.**

(a) Equipment is open to theft and sabotage during the early phases of deployment. Units must provide guards to secure all unit and personal equipment.

(b) Weapons, ammunition, and sensitive items must be stored in a restricted area and guarded.

(c) Classified documents are open to compromise during deployment. Steps must be set up for maintenance, transportation, turn-in, and destruction of classified material. The most secure facility available must be used.

c. **Rear detachment security.** Security after deployment (maintained by the rear detachment) is as important as security before deployment. No information should be given to the news media from other than official sources. The rear detachment may continue to play a role in the unit cover and deception plan.

d. **Tactical operations center security.** Only personnel listed on TOC access rosters are allowed free entry into the TOC area. Personnel wanting to enter the TOC, who are not listed on the access rosters, are detained until their identity and purpose are proven. Such personnel are escorted at all times while within the TOC area.

e. **Marshalling and out-loading area security.** The ranger unit normally marshalls its personnel and equipment in the unit area, providing its own security force. The USAF security police provide this support at the departure airfield. On Army airfields, Army military police provide security. En route security is provided by the RSE.

f. **Personnel and unit equipment sanitization.**

(1) Before deployment, personnel are relieved of non-mission-essential material having intelligence value. This reduces the intelligence available to hostile forces if they capture a ranger force member.

(2) The ranger force commander sets the degree of sanitization IAW mission requirements.

(3) If strict sanitization is imposed, guidelines are issued as follows:

(a) The only personal items retained are identification card and tags.

(b) Unit insignia and markings are removed from uniforms and equipment.

(c) Personal items, such as wallets and photographs, are secured by the rear detachment commander.

(d) Personnel are briefed on how to compose and retain combat diaries. No personal diaries are allowed.

(4) The commander decides how much US currency is to be taken with the unit. '

g. **Area sanitization.** To avoid the compromise of classified or sensitive mission-related material, searches are conducted before and after leaving the host installation, REMAB, or ISB. These searches are held to locate and remove any material, indicators, litter, or other items that could compromise the ranger unit's mission. Unit commanders ensure that a thorough sanitization of their areas is conducted, to include dismantling rehearsal areas, destroying terrain models, erasing chalkboards, and emptying all classified and nonclassified waste containers. During their searches, commanders must pay close attention to emergency operations centers, tactical operations centers, and marshalling and loading areas. The ranger unit's intelligence officer coordinates support from local counterintelligence elements to search the area again after the ranger unit leaves.

3-5. Operations.

a. **Actions upon receipt of the OPORD.**

(1) Task-organize and equip units for special mission(s) IAW the warning order.

(2) Allocate/reallocate mission-essential equipment.

(3) Distribute augmentee personnel.

(4) Redistribute key personnel,

(5) Prepare a deployment time sequence.

(6) Prepare a movement plan.

(7) Prepare and maintain a current operation estimate.

b. **Advance elements.** A planning element is normally deployed to the REMAB/ISB or JTF headquarters in advance of the main body if the ranger force does not deploy directly to the objective area.

c. **Emergency operations center.** An EOC is setup by the ranger battalion headquarters. The EOC is responsible for relaying all reports to higher headquarters. It remains operational until ordered to cease.

d. **Air movement.** The ranger force always deploys by USAF aircraft. This is normally an administrative air movement to a REMAB or ISB. It may be followed by tactical air deployment from there. For some missions, tactical deployment begins at the unit's home station. Even if the movement is administrative, key personnel are cross-loaded. This is done under the direction of the operations section air officer based on aircraft allocations and load plans.

3-6. Logistics.

The following describes logistics readiness needs and steps for deployment.

a. **Supply.**

(1) **Class I.**

(a) Combat rations are normally palletized when the unit is RRF. These rations are to be used during the tactical operation. If the mission calls for deployment directly into the objective area, the palletized stock of rations is off-loaded in the unit area and issued as needed.

(b) Each ranger deploys with emergency combat rations.

(2) **Classes II and IV.**

(a) **Individual clothing and equipment.**

- Individual A bags are packed IAW a predetermined packing list, tagged, and marked. They are palletized, moved to the departure airfield by the RSE, and loaded IAW the air movement plan.
- Individual B bags are packed, assembled, and forwarded on order by the rear detachment.

- Individual environmental equipment is issued. Equipment shortages are replenished out of stocks maintained by the RSE.

(b) **Air items.**

- While the unit is on RRF, a contingency load of main and reserve parachutes is maintained by the RSE.
- Individual and unit air items are palletized by the RSE and shipped in bulk to the REMAB or ISB.
- For direct-deployment missions, air items and packing material are taken to the unit by the RSE.

(c) **General supplies.** The RSE maintains pallets of common contingency items and special equipment. Pallets are picked up by the out-loading support unit and moved to the unit area for issue or moved with the other pallets to the departure airfield.

(d) **Bulk supplies.** Barrier material is identified, relocated, palletized, and out-loaded IAW the air movement plan.

(3) **Class V.**

(a) The unit basic load of ammunition is secured, palletized, and loaded by the RSE IAW the air movement plan.

(b) For direct deployment missions, selected pallets are taken to the unit area and ammunition is issued. The convoy route is coordinated through the post EOC, and secured by the RSE.

(c) Designated items are rigged by the RSE for airdrop, if enough time exists before movement to the REMAB or ISB begins.

(4) **Class VI.** Personnel deploy with comfort items as needed.

(5) **Class VII.**

(a) All weapons and sensitive items not deployed are assembled and secured by the rear detachment personnel.

(b) Unit vehicles are prepared for airland or airdrop.

(c) Inoperable mission-essential equipment is repaired or replaced.

(d) Special equipment requests are forwarded to the logistics section as they are identified.

(6) **Class VIII.**

(a) Individuals deploy with one set of individual Class VIII items.

(b) Unit medics deploy with modified aidbags.

(c) Controlled medical items (less drugs) are palletized and deployed with section supplies. Controlled drugs are secured and issued to unit aidmen on order.

(d) Medical supply chests are maintained in the unit aid station.

(7) **Class IX.**

(a) Units deploy with initial prescribed load list (PLL) items.

(b) Critical PLL shortages are identified to the RSE for follow-on issue.

b. **Maintenance.**

(1) Upon alert, priority maintenance or replacement of key items of equipment is coordinated by the logistics officer. Only minimum mission-essential equipment is replaced.

(2) Items to be returned from supporting maintenance activities are identified, and pickup is coordinated by the property book officer.

c. **Out-loading coordination and support.**

(1) Upon alert, the RSE provides a liaison officer to the ranger logistics section.

(2) The support platoon leader (SPI,) attends the air movement conference and coordinates the following:

(a) Movement times (for both troops and equipment) to the departure airfield.

(b) Steps for weighing and joint inspection of the pallets and vehicles upon arrival at the departure airfield control group (DACG).

(c) Limitations on the pallet height and weight due to aircraft type, allowable cargo load (ACL), and availability.

(d) Changes in the planned number of pallets due to the mission requirements or airframe availability.

(e) Rigging and loading of the heavy drop and supply bundles.

(3) Empty pallets are placed in position for loading by the RSE in coordination with the SPL.

(a) Companies and staff sections prepare pallets.

(b) For airdrop operations, vehicles are prerigged at the unit location. On order, vehicles are driven to the heavy drop rigging site. The SPL and the officer in charge of the rigger detachment coordinate for a rigging site and the delivery of the required air items to that location. Rigging is done by the RSE, and supervised by a member of the ranger unit. Loading aboard the aircraft is IAW the air movement plan and supervised by RSE and USAF aerial port personnel.

(4) The DACG is provided by the host installation. Upon arrival at the DACG area, unit personnel and equipment are under the control of the DACG. As the deployment begins, a

member of the ranger unit is present at the DACG to ensure an orderly loading of personnel and equipment.

(5) The DACG plans for troop-holding areas in the event of delayed or aborted aircraft.

(6) Unit movement data are maintained by the air officer in the operations section.

3-7. Communications-electronics.

a. **Alert notification communications.** Alert notification of the ranger battalions by the regiment is done by secure communications means. Units in local training areas are informed by radio or through their assigned range field telephone. Units operating out of normal FM radio range during RRF periods set up a high frequency net, or other communications, to their higher headquarters.

b. **Telephone service.** Strict control is maintained over key telephones being used during unit alert. Care is taken to ensure that COMSEC and OPSEC violations do not occur. All telephones are assumed to be monitored; classified, sensitive, or official information is not discussed.

c. **Radio communications.**

(1) **Voice communications.**

(a) A secure radio net is set up early in the deployment. The ranger unit EOC, the installation EOC, and the RSE EOC operate in this net. A liaison officer from the ranger unit is placed at the installation EOC.

(b) The ranger support element liaison officer, attached to the logistics section, operates a secure voice station on a separate logistics net to the RSE EOC and the RSE commander.

(2) **Cryptographic material.** During RRF, COMSEC codes and enabling equipment are kept available to set up the secure voice net.

3-8. Remote marshalling base and intermediate staging base.

a. Ranger units can deploy from their CONUS base directly to the objective area. A more common method would be for the ranger unit to deploy first to either a REMAB or an ISB before insertion into the area of operations. (See Figure 3-1.)

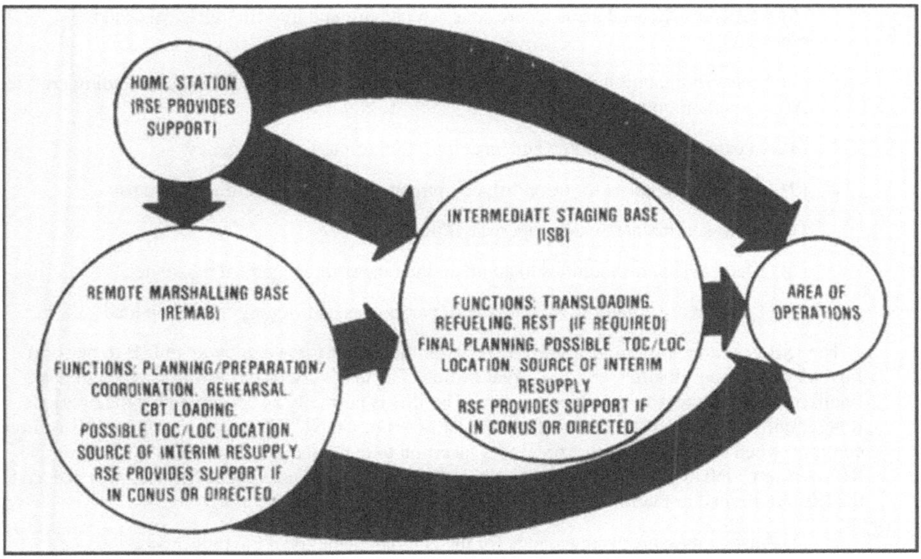

Figure 3-1. Deploying ranger units.

b. The REMAB is a secure base to which the entire ranger unit, including organic and attached support elements, deploys and where mission planning continues. The REMAB is normally within the geographical area encompassed by the command authority of the theater or JTF commander. This ensures that the CSS elements providing support to the ranger unit are operating within their normal area. It prevents or lessens out-of-sector support requirements for CSS elements.

c. The REMAB also provides a secure location for the ranger unit to conduct detailed planning and coordination with the controlling headquarters staff. In the REMAB, the ranger force commander conducts rehearsals, refines and modifies plans, determines PIR, and coordinates with the proper intelligence source to receive that information. It is in the REMAB that specialist augmentees to the ranger force are integrated into the unit, if they have not already joined. Specially trained supporting units, such as aviation and communications elements, also join the ranger force at the REMAB. The REMAB must provide--

(1) Access to the controlling headquarters staff.

(2) Physical security of billeting, planning, maintenance, and communications areas.

(3) Mess, billeting, latrine, and shower facilities for the ranger force and its supporting elements.

(4) Access to a C-141 or C-130 capable airfield, possibly with all-weather operations.

(5) Access to secure communications and processed mission intelligence.

(6) Access to rehearsal areas where sites can be built and live-fire rehearsals can be conducted.

(7) Access to the unit locations of major supporting elements, such as naval landing craft or Army aviation units.

(8) An external security force and an active counterintelligence agency.

(9) Vehicular transport for troop lift, equipment transfer, and administrative use.

(10) Access to maintenance support facilities.

(11) Medical support facilities to augment the ranger force medical personnel.

(12) Covered area to allow for packing of parachutes and rigging of airdrop loads.

d. The ISB is a more austere location. Elements of the ranger force deploy to an ISB to perform final planning, coordination, and task organization. The unit's organization and composition are finalized for movement to the objective area. The ISB is normally not occupied for long periods. Ranger units may deploy to the ISB from the REMAB or CONUS. Deployment to the ISB is more common when terrain or distance precludes insertion to the objective area directly from the REMAB or CONUS. Quick-reaction missions often involve the use of an ISB. Facilities needed at the ISB are limited to the following:

(1) Austere messing arrangements for the combat elements of the ranger force and its attachments.

(2) Austere billets or rest areas.

(3) Water point.

(4) Fuel for aircraft and vehicles.

(5) Areas for test firing of weapons.

(6) Covered and concealed areas for assembly of the ranger force and rigging of parachutes and door bundles.

(7) Austere airfield support facilities, possibly capable of all-weather operations.

(8) Remote location far away from civilians or traffic routes.

(9) Security and counterintelligence elements.

(10) Secure communications.

e. The REMAB and ISB should be located in an area similar in terrain and climate to the objective area. Time spent at the REMAB or ISB lets the ranger force begin acclimatization. Ranger units train yearly in all climates and emphasize physical conditioning. This allows acclimatization quicker than units not as well trained or conditioned. However, ranger units still need an acclimatization period if the terrain and climate of the objective area are different from that of their CONUS base. Not enough acclimatization means that the ranger force could be less efficient, especially when the objective area is hot, dry, or at a great altitude.

CHAPTER 4

Insertion, Extraction, Escape, and Evasion

Insertion is the entrance of a ranger force into a hostile area to execute a particular mission. It is the first critical phase of a ranger operation. These operations often involve deep penetrations of enemy territory by air, water, or land. This chapter emphasizes fundamentals and techniques employed by ranger units when their mission requires insertion and extraction. Extraction is conducted quickly after the mission to avoid casualties.

4-1. Intelligence.

a. Insertion plans are based on timely and accurate intelligence. The headquarters directing the operation must provide current intelligence information to the ranger force. The ranger regiment normally takes action to solicit required information from the controlling headquarters and from other agencies, if security permits. The type of intelligence needed varies depending on the specific mission, but normally includes the following:

(1) Current photographs from aerial or overhead reconnaissance.

(2) Detailed maps.

(3) Data on the population, terrain, and weather from area studies.

(4) Current enemy order of battle, enemy troop locations, and enemy capabilities.

(5) Details on the target area and infiltration routes.

b. To meet its need for current intelligence, the ranger force also relies on aerial reconnaissance and surveillance of the terrain over which it will move. Emphasis is on getting information on the enemy's ability to detect and engage the inserting forces. The location and abilities of air defense radar and weapons systems are critical. Terrain analysis is detailed and focuses on areas suitable for the insertion. Weather information must be detailed and current.

c. In getting information and intelligence data, OPSEC measures are stringent and must be enforced. Coordination with other agencies is kept within security constraints approved by the directing headquarters.

4-2. Deception.

There are always plans to deny the enemy knowledge of the ranger unit's insertion, or to deceive him as to the location or intent of the operation. Feints, false insertions, and other deception operations all add to ranger deception plans. Selection of unexpected means, time, place, and routes of insertion, as well as use of speed and mobility, give less reaction time to the enemy. Planning may include the use of diversionary fires to direct the enemy's attention away from the inserting unit. Deception techniques that may be used include--

a. Multiple airdrops, water landings, or both. This precludes the entire force from becoming engaged at the same time and place, if detected.

b. Dispersion of insertion craft (air or water) both in time and location.

c, Landing a force in an area closer to other potential targets than to the actual targets.

d. Leaks of false information and false messages.

e. False insertions or radio traffic to disguise actual insertions.

f. Diversionary actions such as airstrikes in other than the target areas.

g. Increased reconnaissance flights, either by manned aircraft or remotely piloted vehicles, over false targets.

4-3. Speed and Mobility.

a. Speed of action must be emphasized when the enemy can react quickly to the ranger force presence or when there are limited means of withdrawal. The operation must be conducted swiftly, allowing the ranger force to withdraw before the enemy can react.

b. Major problems facing ranger units are the need for speed and mobility, and the heavy loads carried by soldiers for sustained operations. Ranger units are required to be flexible and to be able to react to any problem within the objective area. However, if equipment is carried to cover every encounter, the load will become too heavy. This will reduce the speed and mobility of the ranger unit. The ranger force commander must always have a balance between the equipment carried and the mobility of his unit. He does this by making three different types of loads: fighting, mission, and existence. He decides what items will make up these loads, where they are carried, and what will be done with them upon contact with the enemy.

(1) The fighting load consists of those items of equipment, weapons, ammunition, food, and water that are common to all soldiers and are necessary for immediate use in combat. These items are normally carried and are retained at all times while in contact with the enemy.

(2) The mission load consists of those items of equipment or ammunition that are required by the mission, but are not normally carried by all rangers in a unit. The load may include spare batteries, demolitions, radios, antitank weapons, or mines. This load is normally spread out among members of the unit and is rotated often. It may be dropped upon enemy contact, but is normally carried to the objective and either used or emplaced there.

(3) The existence load consists of any items designed to provide protection from the elements: sleeping gear, changes of clothes, spare rations, or tentage. This load may or may not be brought into the objective area, depending on the tactical situation. If it is brought, it is normally dropped upon enemy contact or left in an assembly area and retrieved later.

4-4. Stealth.

Methods and steps chosen for insertion must emphasize stealth and try to avoid detection by the enemy. Multiple routes or methods of insertion may be used to preserve the ranger force if some soldiers or units are seen.

4-5. Suppression.

Effort must be made to suppress enemy detection devices, weapons systems, and command and control facilities. This may include electronic countermeasures or direct attack. This helps prevent the enemy from finding and engaging the ranger force en route to the objective area. Deception tactics add to suppression activities. If the ranger force is inserting by infiltration, suppression must not disclose friendly activities.

4-6. Security.

Emphasis on OPSEC prevents compromise of an impending operation. The operation must not be disclosed by overt rehearsals or training activities, the open use or procurement of special items of equipment, or the location of the marshalling area. Several measures can help in maintaining security:

a. Guarded and secure facilities for the headquarters and staff during planning.

b. Issue of operation orders only after units are moved to secure marshalling areas.

c. Alerts, rehearsals, and training to get the local people used to conditions that will exist during actual marshalling.

d. A secure facility set up to isolate units selected for an impending operation.

4-7. Night vision and electronic devices.

a. Rangers use night vision devices to detect and avoid the enemy's forces and his detection devices. These devices also assist in controlling and speeding movement and allow units to traverse almost impassable terrain.

b. The use of electronic homing devices permits precise navigation during reduced visibility. These devices may also be used with visual marking systems.

c. Passive night vision devices are used to assemble rapidly and to reorganize.

d. The use of active night vision devices in assembling, and the degree to which drop zones or assembly areas are "sterilized," depends on the following:

(1) Mission success may depend on not being seen during insertion. When a clandestine insertion is made, the time needed to get to the objective will normally be greater, the use of passive devices will prevail, and the sterilization of the area will be vital. Each ranger must be briefed on sterilization plans and techniques for erasing signs of the insertion.

(2) When detection is likely and the mission depends on getting to the objective before the

enemy can react, speed becomes crucial to the success of the insertion. Insertion may be near or on the objective. Rangers must get in and out quickly. They do minimum sterilization of the area and can use active RSTA devices.

4-8. Rehearsals.

Rehearsals should be close to the actual conditions of insertions and extractions. Security of rehearsals and rehearsal sites is important to the success of the operation. Rehearsals are held on terrain and structures like those in the objective area. If such structures are not available, they are built by the ranger force and the RSE with its supporting engineers.

4-9. Sand tables and terrain models.

Sand tables are useful to acquaint personnel with insertion sites and surrounding terrain. The use of sand tables and terrain models during the issuance of prejump orders and briefings enhances an orderly and speedy tactical assembly.

4-10. Communications.

Reporting during insertion and extraction missions is normally by exception, using an operations schedule.

4-11. Planning insertions.

a. **Centralized planning.** Planning for insertion of ranger units is centralized at the lowest level responsible for the overall operation. This is normally at the joint task force level, with detailed coordination required among all supporting forces. These forces may include other Army elements, USAF, United States Navy (USN), United States Marine Corps (USMC), other US agencies, and allies. Preparation must include detailed contingency planning for use of firepower, shock effect, speed to break contact (if detected), control measures, procedures, and planning. When planning takes a long time, intelligence must be continually updated through all available sources. This includes overhead photography and ground sources to ensure that the initial planning data are still current. Planning stresses the use of deception. Escape and evasion plans, external communication nets, and the location of caches must be known to leaders at all levels.

b. **Reverse planning sequence.** Insertion planning is done in the reverse planning sequence with the ground tactical plan being prepared first. Actions at the objective serve as the basis for the landing and assault plan, the plan for insertion and movement, the loading plan, and the plan for marshalling. All plans are interrelated. Some may be developed at the same time. Consideration of the factors of METT-T guides all planning. Special considerations that apply to these operations are:

(1) **Mission.** The nature of actions at the objective and requirements for special items of equipment and nonorganic personnel impacts on how the unit inserts. When the mission requires rapid deployment, the fastest method of insertion may be needed. In other operations, where success of the mission depends on maintaining secrecy, speed may be less important.

(2) **Enemy situation.** Enemy disposition, strength, and security measures along the route(s) near the objective area affect the method of insertion to be used. Planners must take into consideration whether the enemy:

(a) Uses RSTA detection devices or not.

(b) Has a strong air defense system or possesses air superiority.

(c) Has gaps between his defensive positions.

(3) **Terrain and weather.**

(a) **Topography.** Terrain such as mountains, swamps, or heavy forests favors land infiltration. Deserts, arctic regions, and open prairies favor air or water insertion. Areas not inhabited or with a friendly but scattered population favor land infiltration. A heavily populated area, or one which contains large enemy forces, favors air or water insertion. Hydrography--tides, currents, surf, reefs, and sandbars--must be considered when planning water insertion.

(b) **Weather.** Bad weather with reduced visibility and high winds favors land infiltration, and adversely affects air or water insertion. Limited visibility favors land infiltration; good visibility does not. Adverse weather aerial delivery systems (AWADS) lessens the impact of reduced visibility as a limiting factor on air infiltration. High winds may preclude airborne insertion.

(4) **Troops.** The number of men to be inserted and their level of training may be limiting factors. A need for special skills may call for use of nonranger unit personnel whose physical stamina and ability to perform many infiltration techniques may not equal that of the rangers. Whether supporting forces and special items of equipment are available must also be considered.

(5) **Time.** The main concern is whether the ranger operation is quick-response or deliberate-response. During a quick response mission, planning, preparation, and rehearsal time is short. The timing of the insertion is crucial with respect to weather, the enemy, and if an insertion craft is available.

4-12. Air insertion.

The most rapid form of insertion is by air. Rangers and equipment can be delivered by parachute (static-line or freefall technique), fixed-wing aircraft (airlanded), or helicopter (airlanded, parachute, or rappeling).

a. Suppression of enemy air defense along the insertion corridor may be necessary. The USAF will normally assume the major responsibility for suppressing enemy air defense artillery (ADA) capabilities that may interfere with the inserting forces. This is done by a variety of sophisticated electronic countermeasures applied against enemy equipment and by strikes against known enemy positions. Army fire support elements, special electronic mission aircraft (SEMA), and ground jammers may participate, as well as Navy aircraft, jammers, and gunfire.

(1) A main danger area is the forward edge of the battle area (FEBA) or frontier areas where

the enemy employs most of his sophisticated weapons systems and air defenses. Other dangerous areas are population or troop concentration areas behind enemy lines protected by point air defense systems.

(2) Since the majority of the enemy's detection devices and air defense weapons may be located near the point of entry, fire support, smoke screens (even at night), and suppressive measures may be critical. The use of special items of equipment may be needed to counter the enemy's RSTA effort.

(3) Long-range fires (artillery and naval gunfire) should be planned on known and suspected enemy antiaircraft locations and on prominent terrain features along the route(,s). Once beyond the FEBA, inserting units may be beyond the range of conventional indirect fire. More reliance will then be placed on air assets for fire support and electronic countermeasures.

b. There must be close coordination between the ranger unit and the transporting unit. What the transporting unit does during the insertion has a great impact on the success of the mission. To decrease the chance of detection, maximum use is made of periods of reduced visibility and tactical cover and deception. Drop zones and landing zones are chosen behind tree lines, in small clearings in forests, or on other types of terrain not easily seen.

c. All flight routes over enemy territory should be planned in order to avoid occupied areas and enemy air defenses. Routes should complement the cover and deception plan. They should avoid giving away final ranger objectives.

d. In-flight emergencies, mainly in deep-pedetration flights, should be considered, The ranger force receives a preflight briefing on the route to be flown and is informed of checkpoints along the route. Simple ground assembly plans for contingencies are set up before enplaning. In an emergency, the leader of the unit involved decides whether to continue the mission or abort. This is based on METT-T factors, contingency plans, and distance to the objective as compared to the distance to friendly territory. Contingency provisions for air and water rescue are made.

e. Deciding the DZ location is a joint effort. After receiving input from the air and ground commanders, the joint task force commander makes the final decision. During air assault operations, the ground commander is responsible for the overall process. After reviewing MEYF-T factors, it may be determined that surprise is more important than speed, or vice versa. This may determine the number, size, and locations of DZs and LZs.

f. Planning and preparation continue during the marshalling phase of any operation using air delivery for insertion. A review must be made of items of equipment that are crucial to the ranger force reaching its target area, accomplishing the mission, and returning safely. The need for special supplies and equipment, their transport, the distance to be traversed, the delivery means, and external support are all factors that enhance or reduce the chance for success. Once the means of insertion has been decided, logistics planning continues with emphasis on requesting and coordinating external support. The following must be considered:

(1) Detailed planning for mission logistical needs.

(2) Extra sets of key items of equipment to ensure redundancy and self-sufficiency of

inserting units.

(3) Plans for accompanying, planned, and emergency resupply. The ranger force should prepackage loads to meet specific contingencies. It should consider the use of supplies and equipment already in the operational area.

(4) In quick-response operations, only manportable items of equipment needed to accomplish the mission are carried. Plans and signals must be established for resupply of key items.

(5) Specific plans to evacuate casualties.

(6) All attached personnel and augmentation equipment must be properly prepared for movement.

(7) Provisions must be made for evacuating captured enemy equipment and personnel.

(8) All inserting units should be cross-loaded. Items to be considered in a cross-loading plan are the ground tactical plan, the assembly plan, the configuration of the DZ, the light route and formation of the aircraft, and the number of aircraft available. Cross-loading ensures that key leaders and equipment--for example, door bundles--are evenly distributed throughout the assaulting force. If one or more aircraft abort or are shot down, some key leaders and equipment will still arrive at the DZ. This permits mission accomplishment. Cross-loading aids rapid assembly on the DZ in support of the ground tactical plan. If a ranger is to assemble near the leading edge of a DZ, he is loaded on the aircraft to exit near the front of his stick.

(9) A bump plan is used if an aircraft should abort while still on the ground and there are no spare aircraft. The bump plan ensures that key leaders and equipment are not left behind when the main body takes off. If an aircraft aborts on the ground, key personnel move to another aircraft and replace nonessential personnel. This concept promotes success on the objective.

4-13. Airborne insertion.

a. During the planning phase of an airborne operation, there is no room for error in judgment. It can be planned and conducted well only if there is a constant exchange of information. All personnel involved must be informed of all changes in plans, times, and locations.

b. The reverse planning sequence is the key to the success of an airborne insertion. Most of the planning is based on the ground tactical plan, followed by the air movement plan to support it.

c. The USAF is responsible for flight planning, initial point (IP) selection, and crew procedures throughout the flight. The senior ranger in the aircraft must remain oriented and keep abreast of any last-minute changes. He must coordinate with the aircraft commander.

d. Most airborne insertions are made at either very high altitudes (HALO/HAHO) or very low. For combat operations, drop altitudes may be less than 500 feet above ground level (AGL), depending on the type of parachute used.

e. In airborne insertions, emphasis is on the use of special delivery or navigational techniques since normal insertion is during periods of limited visibility.

(1) **Adverse weather aerial delivery system.**

(a) A self-contained navigational system in some USAF C-130 aircraft is AWADS. The system allows the aircraft crew to fly a specific route to a DZ during darkness or bad weather. The system works with a Doppler radar that determines the ground speed of the aircraft, and an AWADS computer programmed with special flight route information (start point, route, checkpoints, wind speed and direction, and air speed). These two instruments show where the aircraft is in relation to the programmed checkpoints. This eliminates the need for airdrops under visual weather conditions.

(b) Only certain aircraft in formation are equipped with the actual AWADS device. Other aircraft are equipped with station-keeping equipment (SKE). This SKE consists of a computer that tells the aircrew where their aircraft is in relation to other aircraft in the formation.

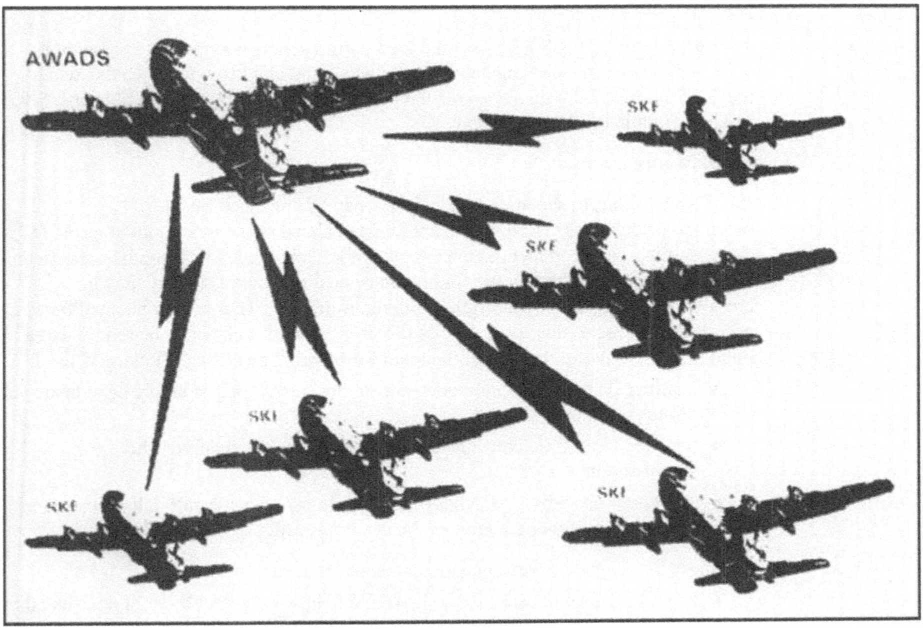

Figure 4-1. AWADS abd SKE aircraft formation.

(c) Airborne insertions using AWADS give the ranger commander the ability to airdrop personnel and equipment during low or zero visibility. This increases the safety of USAF aircraft. It protects them from visually tracked enemy antiaircraft guns and non-all-weather fighters. It also uses adverse weather as a combat force multiplier. Using AWADS, personnel and equipment may be inserted day or night without a pre-positioned USAF combat control team (CCT) or an Army assault team. Time length of the air train is shortened. The air corridor has to be cleared only once. This enhances the element of surprise. The CCTs are brought in with the assault elements. They are able to direct other low altitude parachute extraction systems (LAPES), container delivery systems (CDS), and air-land or heavy drop missions. Use of AWADS enables the ranger commander to conduct rapid, vertical reinforcement of units threatened by an enemy attack during bad weather.

(d) The use of AWADS requires detailed planning, rehearsal, and close coordination among the forces involved. Intense training of both USAF and Army personnel is required.

(e) Personnel must be briefed on terrain through use of photographs, maps, and terrain models. Limited visibility relies mainly on compasses and audible and visual assembly aids.

(f) During reduced visibility, essential heavy-drop serials may precede personnel aircraft. This reduces the time lag between personnel and heavy-drop chalks due to the need to move friendly troops off the DZ. Remaining heavy-drop serials or LAPES can be scheduled following P-hour.

(2) **Reconnaissance teams.**

(a) The operations of the main ranger force may be enhanced by use of reconnaissance teams. These teams are inserted ahead of the main body to provide early warning or to conduct reconnaissance of key objectives. If reconnaissance teams are required at a point along the flight path or near the main DZ, they may be cross-loaded on board any aircraft within the formation. They exit at planned times so as to land as close to their objective as possible. Should such teams be required at a point perpendicular to the axis of the main formation, one of the following is used:

- Another flank aircraft delivers teams to designated points before or with arrival of the main body.
- Selected aircraft deliver the main body, turn, and then deliver the reconnaissance teams.
- The teams are cross-loaded as complete elements on aircraft delivering other units whose flight path crosses the desired locations.

(b) Missions assigned to reconnaissance teams include:

- Providing local security for a USAF CCT, in which case, by joint doctrine, they form the Army assault team (AAT).
- Establishing surveillance or blocking positions on likely avenues of approach into the objective area or DZs.
- Locating and marking targets or assembly areas for the main assault force.

(c) Reconnaissance teams may be inserted by static line or HALO/HAHO operations.

(d) If the reconnaissance element is to locate a specific target or objective, it may be inserted up to 24 hours early. The longer the reconnaissance team remains near the objective area, the greater the danger of it being seen.

(e) In order for the information to be used by the ranger force commander, secure communications and a linkup plan must be prepared and rehearsed.

(3) **Assembly.**

(a) The ranger airborne force can conduct a quick assembly of combat power (weapons, personnel, and command and control elements) on the DZ. Quick assembly and reorganization is crucial because the unit is open to attack. Assembly must be fast and precise. Assembly areas and plans are set forth after considering all the factors of METT-T, mainly the location of the enemy, visibility, terrain, drop information, dispersion pattern, and cross-loading.

(b) The number of assembly areas depends on the location of the DZ, size of the assembly areas, and enemy's detection ability. Unit members move directly to the assembly areas without breaching security. Dissemination of information to all jumpers is the key to a successful assembly plan.

(c) For battalion-size or larger operations, general assembly areas are prescribed in the ranger force OPORD. Company commanders select and designate the exact location of their assembly areas. Although assembly area locations are terrain-dependent, they are generally on the DZ if the drop is at night and off the DZ if the drop is during daylight.

(d) Personnel chosen to carry assembly aids are cross-loaded in the center of their unit sticks. Strong, aggressive leaders should jump first and last in each stick. Upon landing, the leaders roll up their sticks toward the center, locate and move to the proper assembly aid. They may also carry or wear assembly aids. Individual rangers then follow these key personnel. During daylight operations, radiotelephone operators (RATELOS) can display color-coded pennants on their radio antennas to facilitate linkup with leaders. Key personnel may carry and jump with their own radios, and also display color-coded pennants on the radio antennas. Key personnel and RATELOs put their radios into operation before moving to assembly areas. For drops occurring near dawn or dusk, both daylight and night assembly plans are required.

(e) When tactical vehicles or door bundles are dropped, they are normally marked with panels, chemical lights, or strobe lights. The delivery parachutes are often color-coded.

(f) A ranger force should be able to assemble on an unfamiliar DZ, at night, within 45 minutes of the final aircraft pass. Because most ranger operations are time-sensitive, the unit normally moves from the assembly area to the objective when enough of the force has assembled to accomplish the mission. A straggler control point remains at the assembly area under the control of a designated leader. When everyone assembles, or at a specific time, the leader moves to link up with the unit. If available, a radio is left with the straggler control point.

(g) Assembly area security plans are prepared and disseminated. These plans must address hasty defense, sectors of responsibility, observation posts, fire support, and chain of command.

(4) **Contact on the DZ.** During the conduct of parachute insertion, rangers must be ready for enemy engagement at all times. Rangers are open to attack on the DZ before assembling. Just as contingency plans and chance immediate-action drills are developed, briefed, and rehearsed for other chance enemy contacts, rangers must be prepared to execute immediate-action drills for enemy contact on the DZ. The following actions are taken when there is enemy contact on the DZ:

(a) Rangers immediately assemble in fire team-size groups and fight off the DZ to designated assembly areas. Elements attempt to break contact and assemble.

(b) Assembly areas designated by commanders in the mission OPORD serve as primary rally points.

(c) Alternate assembly areas (normally 2 to 5 kilometers from the DZ) are designated in the mission operation order (OPORD). They are used by elements that cannot link up or assemble in the primary assembly area. These elements continue with the mission once assembled.

(d) During limited visibility, personnel should move directly and quickly off the DZ. They must avoid mistaking other friendly elements for enemy, which could cause firefights among friendly forces.

(5) **Reorganization.**

(a) The unit reorganizes according to plans, using designated assembly areas, aids, and identification markings for personnel and equipment. Replacements are named for key leaders or personnel who are lost or injured.

(b) Security teams assemble and move to their positions. Remaining units move quickly to their assembly areas, carrying supplies and equipment.

(c) Subordinate leaders keep the commander informed during assembly and reorganization.

(d) Runners may be used to report assembly status. This maintains radio silence. Designated personnel quickly recover supplies and equipment and move to the assembly areas.

(e) Reorganization is complete when all units have assembled or have been accounted for, and control has been established.

(6) **Pop-up technique.** This involves close terrain flying until the DZ is reached, then the aircraft quickly climbs to jump altitude. Special navigational equipment and aircrew training

are needed but may not always be available.

(7) **HALO/HAHO.** This involves high-altitude, low-opening and high-altitude, high-opening jumps with high performance parachutes designed to maneuver the jumpers to a specific point on the ground. During these operations, midair assembly procedures may be used.

(8) **Arctic.** Airborne insertion in extremely cold conditions is difficult. Times allocated for all parachute and equipment rigging must be increased. Rigging of individual equipment must be IAW FM 57-220. Ahkios can be rigged, packed inside a double A22 container, and dropped in platoon bundles. Because of the increased amount of individual equipment needed, allowable cargo loads for USAF aircraft are reduced.

(9) **Water.** Because of a lack of suitable land DZs, or to avoid detection, rangers may conduct airborne operations onto water DZs close to land. Procedures and techniques are similar to land DZ operations.

(10) **In-flight rigging.** On long-range operations involving extended flight time to the objective area, personnel may rig in flight. In-flight rigging requires detailed planning, coordination, of all jumpers. A rehearsal of these procedures should be conducted before deployment. Aircraft loads must be reduced to carry out in-flight rigging procedures. (See FM 57-220.)

4-14. Air-land insertion (fixed-wing).

a. **Planning factors.** Most of the planning factors that apply to airborne insertion operations also apply to air-land insertion operations. Most USAF aircraft cannot conduct air-land operations in total instrument weather conditions. The AWADS equipment is accurate enough for airdrop navigation, not air-land. Air-land operations are conducted, however, under conditions of extremely reduced visibility. Specially trained USAF crews, using sophisticated navigational techniques, can locate and land on small airfields in bad weather. As in airborne operations, ranger reconnaissance teams, as well as specially trained CCT elements, can jump before the air-land aircraft arrive. They emplace navigational and location markers to assist the pilots. Rangers may rely solely on air-land operations for insertion. This method does not provide for delivery of the force in the shortest time possible. Air-land insertions are normally in concert with an insertion of airborne elements.

b. **Time.** Since turnaround time is crucial, loads in aircraft should be unchained as the aircraft taxies. This also applies to personnel and seat belts. Aircraft should stop as close as possible to the assembly areas where personnel and equipment are off-loaded.

c. **Method.** Air-land insertions normally begin after a ranger Army assault team (AAT) and a USAF CCT have parachuted into the initial objective area and conducted reconnaissance, clearing, marking, and security operations. The ranger ground force commander is the arrival airfield commander. He has control over all arriving, taxiing, and departing aircraft (fixed-wing or rotary) through the CCT, which locates near him. The ranger force command and control element is usually on the first aircraft in and the last aircraft out.

d. **Security.** Air-land operations concentrate large numbers of men and equipment, which presents an open target to the enemy. The ranger ground force commander must ensure rapid off-loading, assembly, dispersion, and movement toward the objective.

e. **Communications.** The ranger ground force commander must have secure radio communications with the ground force, CCT, fire support elements, and higher headquarters. Wire should be laid to connect ground command and control locations.

f. **Operation duration.** Depending on the factors of METT-T, the aircraft may remain on the ground for extraction while the mission is being accomplished or take off at once. They may then orbit, pending recall, or return at a prearranged time.

g. **Extraction.** Extraction of the air-landed ranger force is normally planned and accomplished like an insertion, only in reverse order. A simple, yet effective, way must be found to see that no ranger personnel are left in the objective area.

4-15. Air assault insertion.

Ranger air assault (air-land, rappel, or combination) insertions normally take place during limited visibility. They quickly place the ranger force on, or close to, the objective. The ranger force (depending on the mission) normally plans on fighting when it hits the ground, completing the mission, and then being extracted. Air assault insertions need detailed planning, surprise, flexibility, speed, shock effect, concentration of combat power, and precise timing.

a. **Ground tactical plan.** The reverse planning process is important. The ground tactical plan, developed from the mission assessment, is the first planning area. All other planning begins from this point. The ranger commander and his staff decide stated and implied tasks, and request the proper helicopters for the mission.

b. **Fire support plan.** The fire support available to the ranger commander may be artillery, naval gunfire, mortars, attack helicopters, and USAF or USN tactical aircraft. Because of the timing of ranger air assault operations, preparatory fires are often not used. If they are used, enemy ADA sites normally have priority.

c. **Landing plan.** This plan puts the ranger force into the objective area at the time, place, and proper sequence to support the ground tactical plan. The ranger commander selects primary and alternate LZs and the priority for landing. The exact location of an LZ is determined after a detailed analysis of METT-T. Rehearsals are important. The ranger force must quickly assemble, reorganize, and accomplish the mission. The senior ranger on each aircraft ensures that the personnel on his aircraft offload at the correct LZ.

d. **Air movement plan.** The air movement plan provides for the coordinated movement of the ranger force into the zone of action in support of the 'landing plan. Flight routes, air movement tables, flight formation, altitude, and air speed should be considered. The ranger commander must have secure in-flight communications with his subordinate commanders.

e. **Loading plan.** This plan supports the air movement plan and provides for the pickup zone (PZ) setup and control. It sets priorities for the commander for loading personnel, equipment, supplies, and cross-loading. Air-loading table, aircraft loading formations, bump plan, and rehearsals should be considered.

f. **Staging plan.**

(1) This plan includes PZ selection and operation. The selection of PZs and LZs needs planning and coordination for best use of air assault assets. Site selection in or near the objective must be coordinated in person between the supported ranger commander and the aviation commander. The tactical situation is the key factor in PZ and LZ selection. Other factors are size of landing zones, surface conditions, ground slopes, approach and departure directions, prevailing winds, obstacles, communications, cross-loading, aircraft command and control, PZ and LZ markings, and rehearsals.

(2) The PZs are set up and secured about 15 minutes before touchdown of first aircraft. The fire support plan must support the other plans. All supporting fires must be coordinated with the air mission commander (AMC). Because of the depth of most objective locations, USAF tactical aircraft will be the main fire support method. Orbit locations for fire support aircraft must be planned.

g. **Other planning considerations.** These include escape and evasion, actions at the last LZ, assembly plan, downed aircraft procedures, control measures, forward area rearm and refuel point (FARP), weather delays, and deception plans.

4-16. Amphibious insertion.

Insertion by water may be by SCUBA, surface swimming, small boats, submarines, surface craft, helocasting, or a combination of these methods.

a. **Planning.** Planning for waterborne operations must be thorough. The plan is decided upon between the ranger commander, the transporting unit, and the joint headquarters directing the operation. Plans must be made for all types of enemy action and weather while on board the transporting craft. The transporting unit is given information on a need-to-know basis. Information that could endanger the operation may be withheld until the mission is under way. Initial planning includes:

(1) Time schedule of all events, from the beginning until the end of the joint operation, to be used as a planning guide. Timing for each event must be exact; the success of the operation depends on it. (See Figure 4-2)

Figure 4-2. Waterborne insertion.

(2) Embarkation site, where troops and equipment are loaded on the transporting craft.

(3) Drop site, where the ranger force will depart the primary craft and load into smaller boats.

(4) Handing site, where the unit will beach its boats, or land directly from amphibious craft.

(5) Loading, when loads are lashed and waterproofed IAW unit standing operating procedures (SOPs). Inspections are important. A bump plan for key personnel and equipment must be developed.

(6) Cross-loading, which will help preserve the ranger force. location of leaders, crew-served weapons, and needed equipment are stressed.

b. **Landing site.**

(1) **Beaches.** Beaches are checked to find one that allows an undetected approach. Landing sites that cannot be approached from different directions are avoided. Stretches of defended or patrolled coastline should be avoided. If sand beaches are used, tracks and other signs of

operation that may compromise the mission should be erased. Rural, isolated areas are best. The area behind the landing site should provide an assembly area and concealed routes for exit.

(2) **Barriers.** Reconnaissance may find that good landing sites are not suitable because of barriers between the site and the objective. Such barriers may be patrolled roads, large areas giving little or no concealment, enemy obstacles or defensive positions, areas populated with enemy sympathizers, or an area used for enemy troop maneuvers and training. Unless a means of avoiding such barriers is devised, landing sites should be limited to those without barriers.

(3) **Other factors.** Beach selection should also consider:

(a) The scheme of maneuver.

(b) Surprise.

(c) Enemy dispositions.

(d) Distance to the target.

(e) Landing and exit sites.

(f) Trafficability of the beach.

(g) Cover and concealment.

c. **Routes.** The route the transporting craft takes to the drop site should be planned to deceive the enemy. A major change in the route immediately after the unit's debarkation could compromise the mission. The route should be the same as a route that would be used in some other naval operation--for example, minelaying or sweeping, or a naval patrol. An alternate route must be planned.

d. **Tactical deception.** There should be other plans to deny the enemy knowledge of the insertion besides the water approach route plan. Plans may include electronic countermeasures (ECM), diversions, feints, ruses, or demonstrations.

e. **Actions at the drop site.** Some operations may permit landing directly from the transporting craft. If not, primary and alternate drop sites must be set. The drop site should be at least 1,500 meters offshore to cover any noise in the loading and launching. If the enemy has a surface radar capability, a drop site should be chosen to avoid detection. The drop site may have to be several miles offshore or may need the use of ECM. The manner in which men, small boats, and equipment are loaded must promote fast action at the drop site.

f. **Navigation.**

(1) Ship-to-shore navigation (to the landing site) may be done by dead reckoning. The course may also be maintained by compass navigation, reference to a shoreline silhouette, radar, or prearranged signals from the shore.

(2) Boat handlers, navigators, and other experts are relied on in various phases of the

transport operation.

(3) Lead reconnaissance teams or scout swimmer teams may be used to plant radar, radio, or infrared buoys, or distinct, shielded lights. Boats may be equipped with radar reflectors or beacons to permit accurate tracking.

g. **Actions on the beach.**

(1) Boats hold offshore while ranger scout swimmers move to a covered and concealed position. After reaching the beach, they make a short listening halt. They check the beach and the area immediately beyond the beach for enemy activity. Swimmers signal the force "all clear," mark the landing site, and provide local security.

(2) When the boats land, chosen personnel move directly into covered and concealed security positions. They prepare to defend the landing site.

(3) If the ranger unit is not carrying equipment, boats may be moved off the beach and buried or camouflaged. If heavy equipment makes it hard to move the boats, chosen personnel start unloading at once. The remainder of the force secures the landing site. Each boatload moves into position under the control of a designated leader. Teams are positioned with the best cover, concealment, and fields of fire available.

(4) The ranger force leader sets up a command post from which he can control the operation. He may establish secure communication with his headquarters.

(5) Boats may be deflated and buried or camouflaged near the landing site or away from it. If the boats are to be hidden near the landing site, a team may be chosen to dig holes or cut brush for camouflage along with the unloading of the boats. A small team may be chosen to sweep the beach to erase tracks and drag marks.

(6) The ranger force should be assembled and prepared to move within minutes after the last boat lands. Every ranger must be briefed on the unit's plan should enemy contact be made upon landing.

h. **Insertion by air from ship.**

(1) Helicopters from a ship can launch the ranger force from over the horizon and move at a low level to the LZ.

(2) Helicopters may be vectored from ships to a landing zone. The aircraft may return to the ship while the ranger force conducts the mission, they may orbit offshore, or they may stay at the LZ. Other aspects of landing and assembling are the same as discussed for helicopter insertions.

i. **Helocasting.** This form of insertion combines helicopters and small boats in the same operation. It is planned and conducted the same as air assault operations, except that the LZ is in the water. While helicopters are moving at low levels and low speeds, rangers drop small boats and jump into the water. They then assemble, climb into the boats, and move to shore. Ranger elements are usually assembled and loaded into the small boats within 20 minutes of the last ranger entering the water. This type of operation is normally conducted during limited visibility.

j. **Fire support.** Fire support may be ship-to-air, ship-to-shore, air-to-ground, or ground-to-ground. Specific missions include:

(1) Suppressing enemy air defense.

(2) Neutralizing enemy field artillery within range of boat or helicopter routes.

(3) Employing on-call fires against shoreline targets.

(4) Destroying enemy aircraft.

k. **Contingency planning.** The following must be planned for:

(1) Actions on enemy contact.

(a) En route.

(b) Helocast site.

(c) Flares.

(d) Aerial attack.

(e) Small arms fire.

(f) Indirect fire.

(2) Bump plan with priority to key leaders, crew-served weapons, and radios.

(3) Downed craft procedures.

(4) Escape and evasion.

(5) High surf.

(6) Adverse weather.

(7) Separation.

l. **Rehearsals.** The ranger force must rehearse all aspects of the amphibious insertion. This includes boat launching, paddling, beaching and assembly, boat commands, and capsize drills.

4-17. Land infiltration.

Land infiltration to the objective may be the best way to complete a mission. This is normal when the enemy has total air control or effective air defenses in the operational area to be penetrated, or when the objective is close to friendly territory. Rangers can infiltrate over any type of terrain, in any climate. However, thick forests, swamps, and broken or steep terrain offer the best chance for success.

a. **Planning.** There are many methods of infiltrating by land, and each has its good and bad points. Plans are designed to get the infiltrating force to the objective area without being seen. Plans should include:

(1) Selecting concealed routes for surprise and protection. Primary and alternate routes are based on detailed map and aerial photo study, ground reconnaissance, and data from agent reports.

(2) Avoiding obstacles, populated areas, silhouetting (when forced to cross obstacles or ridge lines), enemy positions, main avenues of approach, and movement along often used routes and trails.

(3) Selecting the time of infiltration to make use of reduced visibility, such as darkness, rain, snow, and fog, and periods of reduced enemy alertness, such as early morning, holidays, and times following protracted movement or combat. The time of the infiltration is important during critical phases (crossing borders or passing through enemy troop concentrations or populated areas).

(4) Using speed and stealth. All rangers must know routes, linkup points, alternate linkup points, time schedules, danger areas, and actions on enemy contact and at linkup points.

(5) Centralizing planning and decentralizing execution. Centralized coordination must ensure that units are acting IAW cover and deception plans. Control should be given to the leaders of the infiltrating groups. Within their designated zones, these leaders select their own routes and formations. Leaders control the rate of advance. When the enemy has electronic detection devices, radio communications must be kept to a minimum. Key leaders normally go with the group that is using the fastest lane or the lane least likely to be detected.

(6) Seeding infiltration routes to the assembly area with unattended ground sensors. These sensors can help the infiltrating force by reducing the chance of surprise contact during the infiltration, and by helping in controlling the forward progress of the units.

(7) Planning for fire support, which must include coverage of routes, danger areas, assembly areas, objectives, and PZs.

(8) Rehearsing actions at the objective, actions on enemy contact, actions at danger areas, and movement techniques.

b. **Movement.** Land infiltration is conducted using one of three techniques:

(1) **Movement in small groups along one axis.** All members of the force use the best route. Small groups are harder to detect, easier to control, and do not compromise the total force if detected. This technique may take too long and it needs an assembly area or linkup point. If the lead group is detected, other groups may be ambushed.

(2) **Movement in one group.** This technique has no reassembly problem. The same route is used by all, making navigation easier. A large force can fight out of a dangerous situation easier than a small one. It has no coordination problems between other infiltrating units. This form of movement is easier for the enemy to detect. If the force is detected, the mission may be endangered.

(3) **Movement in small groups along several routes at the same time.** This technique avoids putting the total force in danger, is less likely to be seen, forces the enemy to react in

many locations, and makes it harder for the enemy to determine the size of the force or its mission. Some groups travel over poorer routes and have problems of assembly, control, and support; the capture of personnel may endanger the mission. Regardless of which technique is used, ranger units normally move to reduce the chance of enemy contact.

c. **Assembly area.** Following movement into the general area of operations, infiltrating units converge on the assembly area or linkup point. Assembly may begin at once after the passage of lines, or it may take place at any time before reaching the objective. The infiltrating force must remain undetected before assembly. If an infiltrating unit is seen at this time, the enemy may learn the location of the assembly area, and the objective. It may take action to destroy the infiltrating force.

d. **Linkup point(s).**

(1) Selection characteristics include terrain that--

(a) Is free of enemy troops.

(b) Offers cover and concealment from enemy observation and patrols.

(c) Does not restrict movement of friendly units.

(d) Accommodates the unit.

(e) Is easily identified.

(f) Offers escape routes to alternate linkup points.

(g) Offers concealment from air attack.

(2) Action sequence for occupying the linkup point.

(a) Halt short of the point.

(b) Establish local security.

(c) Conduct a reconnaissance of the linkup point.

(d) Occupy the linkup point.

(e) Emplace observation posts and listening posts.

(f) Link up with other groups.

(g) Conduct final preparations.

(h) Continue the mission.

(3) Security may be enhanced with the aid of devices such as night vision goggles and night sights. Unattended ground sensors may be placed on likely avenues of approach into the linkup point. Normally, only passive observation devices are used to assist in avoiding detection.

(4) Plans must set a time schedule for actions at the linkup point. All unit members must move in quickly, consistent with security, so that the unit may proceed to the objective rally point (ORP) and on to the objective.

4-18. Stay-behind technique.

A ranger force can be positioned behind enemy lines by employing the stay-behind technique. In this method, a ranger force allows itself to be bypassed as the enemy advances. This is done to allow the ranger force to perform a specific mission. Stay-behind operations may require the concealment, or cache, of supplies before the enemy bypass. Rangers may need to use captured enemy supplies. The location of the assembly area and the plan for extraction are important parts of stay-behind plans.

4-19. Extraction.

The ranger force is normally extracted after the mission is completed. The enemy may also force the unit to abort its mission and to call for extraction. Contingency plans must be made to extract a force under enemy pressure, as well as one that is not in contact. Primary and alternate extraction sites are planned and coordinated with all ,porting forces. Ranger operations are normally conducted against deep targets. Extraction may begin with land movement but normally it includes the use of boats or aircraft. Plans may include primary I alternate means of extraction.

a. **Terrain.**

(1) The terrain is important in choosing the point of extraction - for example, whether or not the extraction site offers good cover from enemy direct-fire weapons and positions from which the extracting force can suppress the enemy. Tidal conditions and PZ size must be considered.

(2) Unlikely locations such as swamps, jungles, or mountain areas are often used for extraction. Specially trained air and naval crews, using the latest extraction devices, make such terrain useable.

b. **Enemy.**

(1) Extraction is best conducted without enemy pressure. However, such pressure could develop during the operation. Detailed plans are made for emergency or contingency extractions forced by the enemy as well as clandestine extractions not under enemy pressure. If under pressure, units will have to use planned fire support to suppress enemy weapons. At the beginning of the extraction with security posted to provide early warning, the unit may start to thin out the same as with a night withdrawal. Ground security may be relied on until the last unit withdraws under cover of suppressive fires.

(2) If enemy contact is made during an extraction, the commander must decide whether to reinsert units or break contact and move to another extraction point. Fire support and suppression are important and include suppressing enemy air defense by deception, ruses, or decoys.

(3) The ranger unit may have special weapons systems or equipment to aid in the extraction. Shoulder-fired antiaircraft missiles may be used to defend against enemy aircraft threatening

the extraction.

c. **Extraction by air or water.** Extraction by air or water is preferred when the resources are available and their use will not endanger the mission. Other factors that favor their use are when--

(1) Long distances must be covered.

(2) Time of return is essential.

(3) The enemy does not have air and naval superiority.

(4) Heavily populated hostile areas block land movement.

(5) Ranger forces are burdened with prisoners of war,casualties, or critical items of equipment.

d. **Extraction by land exfiltration.** Stealth and evasiveness are key elements of exfiltration. This method of extraction is favored when--

(1) The enemy has air superiority.

(2) The enemy is able to prohibit air or water extraction.

(3) The distance to friendly lines is short.

(4) The terrain provides cover and concealment (for movement on foot) and limits enemy mobile units.

(a) Multiple routes of exfiltration may be used to prevent destruction of the force if soldiers or units are detected. Captured enemy vehicles and equipment may be used to assist in the exfiltration.

(b) The ranger force may exfiltrate in one body or in small groups. Exfiltrating in small groups avoids the time delay of assembly.

(5) The exfiltrating force is lightly equipped and not burdened with captured personnel or material.

(6) The exfiltration route passes through an area occupied by friendly civilians or guerrilla forces who can assist the movement.

(7) Areas along exfiltration routes are not inhabited.

(8) The enemy force is dispersed or is not able to concentrate against the exfiltrating force.

4-20. Escape and evasion.

a. A vital part of all premission planning is the development of en-route and post mission E&E plans. Such plans enhance survival of the ranger force, and the transporting air or boat crews. The ranger force commander is responsible for--

(1) Helping in the development of an E&E plan in coordination with all supporting agencies.

(2) Ensuring that all members of the ranger force and the supporting boat or air crew are briefed on the E&E plan.

b. Each E&E plan is unique. Each situation has unique problems associated with E&E. The E&E plan devised by the ranger force commander must address these problems, while profiting from the individual abilities and training of the rangers and their supporting air and boat crews. The following considerations apply to E&E plans devised during ranger operations:

(1) The plans enhance survival of personnel who can no longer accomplish their assigned mission. The senior ranger in a group must decide if there are any remaining missions that the group can accomplish. If not, then he must try to evade and escape enemy capture if unable to link up and be extracted with the rest of the ranger force.

(2) Because of the depth of penetration behind enemy lines, most successful E&E plans may involve either air or water movement away from enemy-held territory. The air and boat crews supporting ranger operations are highly skilled and dedicated. If there is any chance of a successful rescue, these forces continue to try to retrieve rangers escaping from an enemy-held area.

(3) Rangers possess stamina and endurance. They can move a great distance (especially at night) and over rugged terrain to reach an area where they can rendezvous with search-and-rescue (SAR) aircraft or boats. Rangers are also resourceful and highly trained in communication, stealth, and camouflage. This, coupled with their self-discipline, means that successful E&E can be conducted over a great distance, through heavy concentrations of enemy, and over a long period.

(4) Escape and evasion plans for ranger elements should include avoiding contact with the local civilians. However, the aid of friendly insurgent forces may be enlisted. The ranger E&E plan may also include the use of other special operations forces E&E networks that are in place behind enemy lines. These nets must not be compromised by the volume of evaders.

c. Escape and evasion plans should be based on three phases:

PHASE ONE: That portion of the insertion following entry into the hostile zone where personnel survival and recovery is the only thing considered. Should an aircraft or boat carrying rangers be disabled during phase one, the ranking survivor assumes command. He must consider E&E planning and the experience and skills of other survivors. He should be concerned with the security of the mission, caring for wounded, and moving away from enemy activity while contacting SAR elements for recovery.

PHASE TWO: That portion of the insertion that is near the objective and could permit the ranger element to successfully pursue its mission. Although part of the E&E plan, actions by the ranger force during this phase become a forced land infiltration rather than an E&E. The senior surviving ranger must decide if enough personnel and equipment can be moved to the objective area in time to accomplish the mission. If so, he must proceed, taking along any surviving air or boat crew members. If not, he may proceed directly to the extraction point and wait, or he may contact SAR elements for recovery.

PHASE THREE: That portion of the operation after mission accomplishment. It is a hard time to escape and evade since the enemy is alert and reacting to the ranger forces. The confusion caused by the violence of the attack, coupled with the rangers' ability to move quickly cross-country away from the objective area, offers the best chance of a successful escape and evasion. The surviving ranger elements that cannot be extracted can do one of the following:

(1) Move over land at once to a PZ or beach for extraction by SAR elements.

(2) Move to an area where reentry into friendly lines can be made.

(3) Move to a neutral or nonaligned country where either internment or return to friendly control can be effected.

d. Any plan for E&E should provide for the issue to each ranger of an area map, a covert means of recognition, and a small compass. An emergency ration may also be included and any items needed for the climate in that objective area. The ranger force commander must balance the need for other survival equipment against the added weight and bulk involved.

CHAPTER 5
Strike Operations

The ranger regiment conducts strike operations in support of the Air Land Battle. Ranger strike operations are integrated into an overall plan designed to destroy, delay, and disorganize the enemy. They may also cause him to divert a major part of his combat forces and power to rear area security. Strike operations also help create an environment in which other special operations forces can operate. Strike operations can complement air power in disrupting enemy forces in depth. They allow the theater commander to regain the initiative and lead to decisive action.

5-1. Application.

a. The ranger regiment can conduct strike operations either as deliberate-response or quick-reaction missions. It can be inserted into the area of operations by land, sea, or air.

b. Strike operations can be conducted throughout the depth of the battlefield in support of conventional or unconventional warfare. Multiple strike operations can have a synergistic effect and can create situations that help friendly forces seize and maintain the initiative.

c. Strike operations include raids, personnel and equipment recovery, and interdiction of lines of communication. They are best accomplished by ranger units who use conventional techniques and special skills. Strike operations vary in size. When well augmented and supported, the ranger regiment can conduct strike operations deep in enemy territory using up to three battalions. Strike operations normally are limited in time and may be conducted unilaterally or in conjunction with other special operations forces or allied forces. They are normally ended by the extraction or exfiltration of the ranger force.

5-2. Raid operations.

a. A raid is a strike operation conducted behind enemy lines against strategic objectives, targets of high tactical value, time-sensitive targets, or key personnel and facilities in enemy rear areas. It normally involves a swift penetration of hostile territory to confuse the enemy or destroy his installations. Ranger units can conduct deep penetration raids when given the appropriate assets for insertion.

b. A strategic raid by the ranger regiment would be directed by the NCA and conducted under the OPCOM of a unified or JTF commander. An operational raid would be directed by the unified or JTF commander and conducted by that command directly or under the OPCOM of a corps commander. Typical targets would include:

(1) Command, control, communications, and intelligence centers at front level or combined arms army level.

(2) Nuclear and chemical weapons storage sites and delivery means.

(3) Key logistical centers, warehouses, ammunition complexes, or fuel pumping centers supplying logistical support to a specific front.

(4) Air defense and air traffic control integrating centers and air defense weapons sites.

(5) Key power generating and transmitting facilities, hydroelectric dams, and irrigation systems.

(6) Key ports or rail complexes.

(7) Key installations or facilities, such as airstrips, buildings, bridges, dams, or tunnels.

c. The size of the ranger raid force is determined by a detailed analysis of METT-T. The unit committed to the ground phase of the raid mission is kept as small as possible. It may be as small as a ranger company or it could include all the battalions of the ranger regiment. The force must be tailored to complete the mission quickly, violently, and with few casualties. The size of the force may also be affected by the augmentation of specially trained personnel needed for that mission.

d. Raids are normally conducted in five phases. (See Figure 5-1).

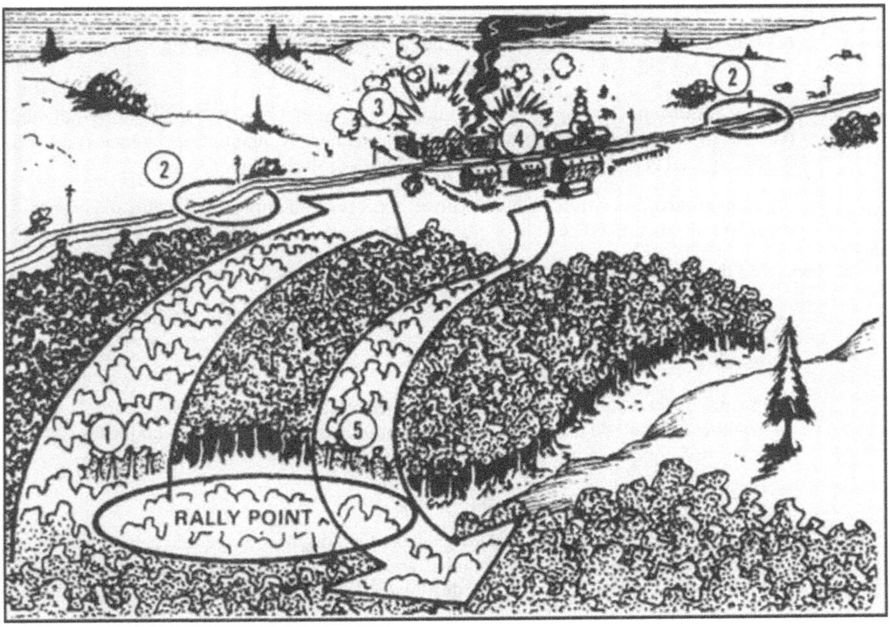

Figure 5-1. Five phases of a raid.

PHASE 1. The ranger force inserts or infiltrates into the objective area.

PHASE 2. The objective area is then sealed off from outside support or reinforcement, to include the enemy air threat.

PHASE 3. Any enemy force at or near the objective is overcome by surprise and violent attack, using all available firepower for shock effect.

PHASE 4. The mission is accomplished quickly before any surviving enemy can recover or be reinforced.

PHASE 5. The ranger force quickly withdraws from the objective area and is extracted.

e. This paragraph discusses the raid after insertion/infiltration. All considerations discussed in Chapter 4 must be applied in order for the unit to get deep behind enemy lines. Upon receipt of a raid mission, the following basics are emphasized:

(1) **Maximum use of intelligence information.** The gathering and dissemination of information must be continuous and provided to the raid force even while en route to the target area. To ensure mission accomplishment, the ranger unit must be kept informed of the latest enemy developments in the objective area to prevent being surprised. Only then can the full combat power of the rangers be concentrated at the decisive time and place.

(2) **Plan development.** The reverse planning sequence and the planning steps set forth in FM 101-5 are followed in the plan development.

(3) **Coordination.** Coordination is through the operational headquarters with friendly units (such as higher and supporting Army or joint headquarters, intelligence agencies) for fire, RSTA, special equipment, and personnel and logistical support.

(4) **Rehearsals.** Rehearsals help validate all aspects of planning for the raid and ensure precision in execution-They allow changes to be made in the plan before it is carried out.

f. A successful raid is ensured by--

(1) Launching the raid at an unexpected time or place by taking advantage of darkness and other periods of limited visibility, and moving over terrain that the enemy may consider impassable.

(2) Avoiding detection through proper movement techniques and skillful camouflage and concealment to include taking advantage of natural cover of the terrain, and using sophisticated equipment to detect and avoid enemy forces.

(3) Timing the operation as close as possible.

(4) Using all available support, both organic and nonorganic, to include use of special weapons such as Air Force "smart bombs" and artillery cannon-launched guided projectiles, with the ranger unit using laser target designators.

(5) Performing quick, violent, precise, and audacious actions that focus full combat power at the decisive time and place.

(6) Disengaging quickly upon mission completion.

(7) Withdrawing swiftly using planned routes and including a deception plan.

g. Four functions are normally performed by ranger tactical formations when conducting a raid. Each subelement is organized and equipped to do a specific part of the overall mission. Depending upon the specific mission, nature of the target, enemy situation, and terrain, these four functions are as follows:

(1) The command group controls movement to and actions at the objective. This unit normally consists of the ground commander, other subordinate leaders in the raid organization, and communications to support these leaders.

(2) The security element, whose organization is determined by the mission of the raid force, size and type of enemy force and its mobility and state of alert, terrain and avenues of approach into the area, and the time needed to seal off the objective area. The security element may perform the following:

(a) Secure the ORP.

(b) Give early warning of enemy approach.

(c) Block avenues of approach into the objective areas.

(d) Prevent enemy escape from the objective.

(e) Provide overmatch for the units at the objective and suppressive fires for their withdrawal.

(f) Provide short-range air defense.

(3) The support element provides the heavy volume of fire needed to neutralize the objective. Because fires from this unit are violent and devastating, they must be closely controlled to ensure the precision needed. On order or as planned, fires are lifted and shifted to cover the maneuver of the assault element by suppressing enemy fire from the objective or aerial fires. The support element may also be given specific locations to cover by fire in support of the security element if an enemy quick-reaction force moves toward the objective area. These may include routes into and out of the objective site, key terrain features, or installations adjacent to the main objective. Once the assault has been completed, or on order from the raid force commander, the support element displaces to the next planned position. Organization of the support element is determined by the following:

(a) Size of the objective, the geography of surrounding area, and the enemy threat (to include air) in the area. This element should be able to neutralize the objective (when supported by air or naval gunfire) and to lift or shift fires either when the assault is launched or when ordered by the raid force commander.

(b) Mission of the assault unit.

(e) Suitable firing positions.

(d) Size and nature of the enemy force in the objective area and those enemy forces capable of reinforcement at the objective.

(e) Fire support from other units (air strikes, naval gunfire, surface-to-surface missiles, and artillery fire).

(4) The assault element seizes and secures the objective and protects demolition teams, search teams, prisoner-of-way teams, or other teams.

(a) The organization of the assault element is always tailored to the mission. Each target area must be examined carefully. To assault, seize, and destroy an installation, position, or equipment, the assault element could be organized into one or more assault teams. The element's mission is to overcome resistance and secure the objective and to destroy the installation or equipment. Other specialized teams may also be needed. For example, sniper teams could be needed to remove key sentries. To capture prisoners, liberate personnel, and seize or destroy equipment, the assault element could be organized into assault teams, prisoner teams, search teams, medical teams, demolition teams, or breach teams.

(b) To destroy a point target or installation in a heavily defended area where the USAF cannot get close en to be effective, the assault element might be organized with one small team equipped with laser target designators. From covered and concealed positions, members of the team could then provide guidance for USAF delivery of laser-guided munitions from a safe distance.

h. The site chosen for the raid LZ or DZ must support the planned actions at the objective. There are two options when choosing sites:

(1) The ranger force can land on or near the objective and seize it before the enemy can react. (See Figure 5-2.) This avoids forced marches over land carrying heavy combat loads. If there is no suitable landing area near the objective, or the enemy has a strong reaction force nearby, this option is not favored.

Figure 5-2. Airborne landing on the objective.

(2) The ranger force can land unseen far from the objective. It then assembles, reorganizes, and moves into an ORP near the objective. (See Figure 5-3.) The objective is seized after security and support elements are in place. This option makes coordinated action easier by setting up control of small units before engaging the enemy.

Figure 5-3. Airborne landing distant from the objective.

5-3. Interdiction operations.

a. AirLand Battle doctrine set forth in FM 100-5 is a shift in emphasis toward keeping the initiative in battle. The deep battle seeks out enemy elements separated from the main battle area and attacks them before they can be deployed. Interdiction of selected enemy forces by attacking them or blocking chokepoints hinders, delays, or interrupts the enemy's use of his lines of communication; it denies him the use of certain areas; it destroys industrial facilities, military installations, and equipment; and it stops or slows his advance along a particular route of approach. The ranger regiment's ability to conduct interdiction is vital to execution of the AirLand Battle doctrine.

b. Interdiction by ranger forces includes blocking or channeling approaches, holding key terrain for a limited time. and destroying facilities.

c. Ranger forces conduct interdiction missions that supplement those conducted by long-range Army and USAF weapon systems. Large-size ranger forces are normally used only when other systems fail or are not available, and the target is of such priority that it must be attacked. Small ranger laser target designator teams may work in concert with USAF or USN strike aircraft to direct terminally guided weapons systems. Rangers should be directed against crucial interdiction targets that disrupt, confuse, or delay the enemy and allow friendly forces to gain the initiative at the forward line of own troops (FLOT).

d. Typical interdiction operations are--

(1) Destroying rail lines, switches, rail yards, and rolling stock.

(2) Destroying bridges, overpasses, tunnels, and truck parks.

(3) Closing mountain passes or routes in restricted terrain.

(4) Cutting oil or natural gas pipelines and destroying pumping and distribution systems.

(5) Destroying dams, locks, spillways, and flood control gates.

(6) Destroying rail transload points, switches, and yard engines.

(7) Destroying shipyards, dry docks, wharfs, lighters, cranes, and stevedore facilities.

(8) Destroying electrical generation facilities, power lines, transformers, and grid monitoring centers.

(9) Destroying water pumping stations and filtration systems.

(10) Destroying radio and TV stations, phone lines, microwave terminals, satellite receiving stations, and telephone exchanges.

(11) Ambushing enemy forces and personnel moving along lines of communication.

e. Considerations when using rangers for interdiction are:

(1) Insertion and extraction.

(2) Support resources available.

(3) Ranger force vulnerability.

(4) Target identification.

(5) Threat environment.

(6) Other or supporting means of interdiction.

(7) Command and control.

(8) Time and magnitude of the effect on the enemy forces at the forward edge of the battle area.

(9) Collateral effects of target destruction.

f. Interdiction targets are not identified and attacked at random. They are part of an overall deep attack plan to destroy or cripple an entire system. Interdiction is based on the assigned deep campaign plan that directs which targets are to be attacked, by whom, the results desired, and the priorities of attack for specific systems. Based on his plan, the ranger force commander selects those ranger elements best suited to conduct the attack. Target selection, which requires detailed intelligence and planning, is based on:

(1) **Target criticality.** A target is critical when its destruction or damage would have a major influence on the enemy's ability to conduct or support operations. Each target is considered in relation to other elements of the particular target system designated for interdiction. The criticality of a target changes with the situation. For example, when the enemy has few locomotives, a railroad bridge is less critical.

(2) **Target accessibility.** The ranger force must be able to insert into the target area. Terrain, security controls, location, and means of insertion are considered when studying a target for accessibility. systems. Based on his plan, the ranger force commander selects those ranger elements best suited to

(3) **Target restorability.** The enemy's ability to restore a damaged facility to normal operating capacity affects target selection. The enemy's repair and maintenance ability and his reserve stocks must be considered.

(4) **Target vulnerability.** A target is vulnerable when it is open to attack by means available to ranger forces. Vulnerability is affected by the nature of the target such as type, size, disposition, and security. It is also affected by the means available to attack it such as explosives, incendiaries, and special devices.

g. Key to the interdiction process is the ability to locate and evaluate target systems for attack and the timing of the attack. A target system is a series of related functions that serve a common purpose.

(1) A target is one element, installation, or activity identified for attack such as a locomotive, train, bridge, or prison.

(2) A target complex is several targets in the same general area, such as a railway marshalling yard, an airfield, or port and dock facilities.

(3) A target system consists of an industrial system and its sources of raw materials; the rail, highway, waterway, or airway systems over which these materials are transported; the source of power and method of transmission; the factory complex; and the means by which the finished product is transported to the user. The targeting process is usually accomplished in the special operations force cell at EAC and the battle coordination element (BCE) at corps level.

h. Interdiction techniques include ambush, emplacement of conventional demolitions, target designation for laser-guided munitions, and emplacement of beacons or sensors.

NOTE: The enemy surface-to-air threat normally precludes the use of an airborne platform for direct observation and designation. This may be accomplished by the on-ground ranger force.

i. Through interdiction, ranger units are an ideal force to contribute to the Army's AirLand Battle operational concept. This includes taking the initiative, moving quickly to strike deep in the rear of the enemy, and defeating the enemy force beyond the FEBA. Planners and commanders must ensure that the ranger units are used on missions that are synchronized to contribute to the overall deep battle campaign. Strikes against critical targets must combine with efforts of other assets to take the initiative from the enemy, limiting his flexibility, initiative, and momentum.

5-4. Personnel and equipment recovery operations.

Recovery operations are specialized raids to liberate imprisoned or detained personnel or to return certain items of equipment to friendly control. These operations include recovering and extracting downed or hijacked aircrews and political or military leaders. Rangers use surprise and combat power to overwhelm resistance before harm can come to the detainees or prisoners.

a. **Planning.** The planning of personnel and equipment recovery operations is the same as for the raid. The main difference is that the ranger force commander must plan for the extraction of the recovered personnel and the loading and extraction of sensitive equipment. Personnel and equipment recovery raids are often executed as quick-response missions. They normally require the ranger force commander to plan and execute the recovery quickly.

b. **Augmentations.** This type of strike operation often requires that the ranger force be augmented with nonranger personnel with special skills or abilities. Examples are medical personnel, technical experts, mechanics or crew members trained in repair and retrieval operations, linguists or translators, and other special operations force elements.

c. **Organization.** The organization of the recovery force is the same as for a raid. Sometimes special teams are needed to perform certain missions involved in the recovery. This mission is conducted aggressively, usually ending in close combat during darkness or other limited visibility conditions. Reconnaissance teams may be inserted ahead of the ranger force to reconnoiter the objective and to locate the detained personnel.

(1) All planning and execution takes place as described for the raid. The emphasis must be on detailed, timely intelligence.

(2) Medical teams must be available to care for the detainees.

(3) The ranger personnel must be adept at searching and clearing of buildings.

(4) The ranger force must--

(a) Locate the specific building or area where the friendly detainees are being held.

(b) Gain entry into buildings, use stun grenades, and avoid obstacles and booby traps.

(c) Engage all enemy personnel using selective-kill techniques.

(d) Locate, identify, and secure detained personnel.

(e) Move personnel to a secured extraction site.

(f) Leave no detainees or rangers behind in the objective area.

d. **Execution.** The key to success is surprise and speed. The enemy must be assaulted when least expected. The ranger force must ensure that friendly detainees are not harmed during the assault - either by rangers or enemy personnel.

e. **Extraction.** The ranger force commander should plan for the extraction point to be as close as possible to the point of recovery. This prevents the detainees from having to move cross-country or for the ranger force to transport bulky equipment long distances. For small groups of detainees or small items of equipment, the extraction point can be further away. Recovery operations may use any method or combination of methods of insertion and extraction (see Chapter 4). Close planning and coordination is normally required with US Army, USAF, or USN aviation for both delivery and evacuation of the target area.

CHAPTER 6

Special Light Infantry Operations

Ranger units are normally employed on strike operations. However, broad US strategic needs demand a wide range of other military abilities. The ranger regiment may conduct special light infantry operations when larger conventional forces could not be used. The regiment conducts these operations using light infantry doctrine, methods, and techniques.

These operations include, but are not limited to, movement on foot, military operations on urbanized terrain (MOUT), passage of lines, ambush, and reconnaissance. Ranger units do not normally conduct planned operations against enemy armored forces. Battlefield surprises or METT-T considerations may cause ranger units to engage armored forces for short periods. The ranger regiment can conduct a conventional light infantry defense for short periods. A defense is not normally conducted adjacent to other friendly forces. It is normally a perimeter defense from which the unit then breaks out from encirclement and links up with friendly forces.

6-1. Movement.

a. Most missions involve movement on foot in tactical operations. Ranger units infiltrating behind enemy lines seek to avoid chance enemy contact. They choose the movement that allows them to retain security and control. To avoid loss of surprise and initiative, casualties, and mission failure, ranger units normally--

(1) Avoid chance enemy contact, if possible.

(2) Move on covered and concealed routes.

(3) Avoid likely ambush sites and other danger areas.

(4) Practice camouflage, noise, and light discipline.

(5) Maintain all-round security.

(6) Make unavoidable enemy contact with the smallest element possible.

(7) Retain the initiative to attack at the time and place of the ranger unit's choice.

b. In planning the movement of his unit, the ranger commander must consider all the elements of METT-T and determine how they will affect the movement. He must determine the route, formations, security measures, methods of command and control, location of key leaders and weapons, and action to be taken on chance enemy contact.

(1) **Routes.** The ranger force commander conducts a careful route analysis, using the factors of observation, concealment, obstacles, key terrain, and avenues of approach (OCOKA). He chooses a route that maintains the security of the force while ensuring surprise. He tries to choose a route that will avoid enemy contact and speed movement. The commander makes a detailed route reconnaissance using maps, aerial photographs, and reports from persons familiar with the area. If possible, he reconnoiters the route himself or uses a reconnaissance team. He may have that team act as guides or mark the route using covert marking devices.

(2) **Formations.** Ranger units use the movement techniques found in current light infantry doctrinal publications. The traveling, traveling overmatch, or bounding overmatch technique is chosen based on the likelihood of enemy contact. The ranger force normally tries to avoid detection and contact with the enemy while moving. The commander chooses a movement technique that ensures that if unexpected contact were made, it would be with the smallest element possible and not the entire formation. The distance between moving ranger elements depends on the terrain, visibility, and enemy situation. March interval is normally great enough to allow each succeeding element either to deploy or change direction if enemy contact is made. Distances are close enough that each element could quickly assist the element to its front.

(3) **Security.** Security is to the front, rear, and flanks when enemy contact is likely. A front security element is placed well forward with adequate communications with the main body. The security element is used to detect the enemy and warn the main body. The main body then either changes direction or hides until the enemy is past. If the enemy does detect the main body, the security element may be used to attack the enemy from the flanks or rear. Flank security elements are placed next to the route of march. They move forward either by alternate or successive bounds. An alternative is to have the security element moving adjacent to the column along routes paralleling the direction of march. Rear security is handled the same as front security. Plans are made for the rear guard to support the main body during chance enemy contact. This is done either by maneuver or by furnishing supporting fire.

(4) **Command, control, and communications.** All available communication is used, consistent with OPSEC and movement security, to assist in maintaining control during movement. March objectives, checkpoints, and phase lines may be used to aid in control. The number of reports is reduced, normally only exception reports are needed. The ranger force commander should be well forward in the formation but may move throughout as the situation demands. Communications with security elements are mandatory. Operations security often precludes the use of radios. Therefore, connecting files, runners, and visual signals are normally used. Detailed planning, briefing, rehearsals, and control are valuable if there is enemy contact. Alternate plans are made to cover all possible situations.

(5) **Weapons.** Rangers are placed within the formation to allow a mix of their individual weapons to cover the front, flanks, and rear of the formation. Rifles, grenade launchers, and light automatic weapons cover all sectors. Crew-served weapons such as machine guns, antitank weapons, and mortars should be given designated locations within the formation. This allows the element leader to employ them quickly. They should be placed near the lead element, but not so close as to be engaged along with it. Crew-served weapons are used to

provide overmatch, especially when employing the bounding overmatch technique of movement.

(6) **Rehearsals.** The ranger force, expecting movement as part of an operation, conducts the following combat drill rehearsals:

(a) Conduct movement.

(b) React to enemy contact.

(c) Maneuver under fire (fire and movement).

(d) Cross danger areas.

(e) Break contact.

(f) Conduct a hasty ambush.

(g) Make a temporary halt.

(h) Establish a perimeter.

(i) Establish a rally point.

(j) Take actions upon consolidation.

(k) Use arm-and-hand signals.

(l) React to flares and indirect fire.

c. After a ranger force is inside enemy territory, it must be alert to avoid detection before reaching the objective.

(1) Should the ranger force security element become aware of enemy presence, it alerts the main force. The main force tries to move away without alerting the enemy and continues the mission.

(2) If the enemy does make contact, the security element then engages the enemy, trying to deceive him as to the size of the ranger force and its direction of movement. After breaking contact, the security force divides into small groups and moves to a rally point to rejoin the main force.

(3) The ranger force does not fight en route to the objective unless there is no alternative, then it quickly breaks contact and moves away.

(4) Ranger units may need to set up a temporary defensive position for resupply, evacuation of wounded, planning, or extraction. One technique that can be used is the temporary perimeter defense.

(5) If the moving ranger force comes under indirect fire, it moves quickly out of the impact area and continues the mission.

(6) The most effective means of combating an ambush is through the immediate return of a

high volume of fire by all weapons. This requires alertness, discipline, dynamic leadership, and rehearsed plans.

(a) Regardless of the method of movement, all weapons should be positioned for immediate use. A well-executed ambush is hard to counteract. However, the effectiveness of an ambush can be reduced by the immediate return of fire. Enemy weapons may be silenced, thus creating weak points in the ambush position. Smoke and high-explosive (HE) grenades are effective in executing counterambush actions. They create confusion, provide a screen for movement, and disrupt the ambush plan.

(b) When all elements of a ranger force are not trapped within the ambush, the elements that are free to maneuver should initiate an immediate flank or rear assault against the ambush force. A flank assault permits better coordination with personnel in the ambush. Supporting fires are provided by those weapons that cannot be hand carried and fired in the assault.

(c) Specific actions against a near and far ambush are as follows:

- **Near ambush.** Elements caught in the kill zone return fire at once and move out of the kill zone. Elements not in the kill zone help by maneuvering and conducting a flanking attack. The unit assaults the ambush position.
- **Far ambush.** Those ranger elements not in the kill zone set up an overmatch position and lay down a base of fire for the elements in contact. The elements in contact move out of the kill zone and prepare to assault the enemy, using individual movement techniques and fire and movement.

(7) The reorganization after enemy contact involves the use of rally points, plans for local security, reorganization of the force, evacuation of dead and wounded, and movement based on the unit mission.

6-2. Ambush operations.

a. **Definition.** An ambush is a surprise attack, from a concealed position. It is used on a moving or temporarily halted enemy to destroy or capture him and his equipment. An ambush is a useful tactic because--

(1) Small, well-trained, disciplined forces, with limited weapons and equipment, can destroy much larger enemy forces.

(2) It reduces the enemy's overall combat effectiveness by destruction and harassment of his forces.

(3) Enemy morale and effectiveness suffer heavily at little cost to the force executing the ambush.

b. **Execution.** A successful ambush must be executed with precision, violence, speed, and audacity of execution. For success, ambush operations must emphasize the following:

(1) **Surprise.** Surprise, more than any other single aspect, enhances the value of an ambush. Surprise increases the potential for inflicting damage on the enemy with less danger to the

ranger force.

(2) **Coordinated firepower and shock effect.**

(a) Coordinated firepower is used for maximum shock effect. Massive volumes of accurate fire, explosives, and mines, coupled with an aggressive attack, breaks the enemy's spirit to fight back. Surprise increases shock effect and the chances for success. Shock effect can cover unexpected defects in an ambush--for example, ambushing a much larger force than expected.

(b) All weapons must be sited with interlocking fires in the kill zone and along likely avenues of entrance or exit. Mortars should be used if the terrain permits. Tripods and traversing and elevating mechanisms are normally used with machine guns to lock in fires. All riflemen use firing stakes to mark left and right limits, and elevation stakes. There is a tendency to shoot high in an ambush--especially at night. The M203 grenade launchers are sited to cover the dead space and routes of escape. When the rules of engagement permit, plastic bags containing CS crystals may be taped to the front side of Claymore mines.

(3) **Control.** Control is essential.

(a) Leaders must have contact with all members of their unit to alert them to the oncoming enemy. Leaders should not move around the ambush site during this crucial period. A method used to alert members can be to tie strings or vines to rangers' legs or arms. By a series of light tugs, all members of the ambush can be alerted to enemy presence.

(b) The leader must initiate the ambush with a casualty-producing device. A bank of Claymore mines on a double-ring main is an excellent device to spring an ambush. Other good techniques are to use a 90-mm recoilless rifle firing antipersonnel (APERS) rounds or a machine gun. All must be under the direct control of the ranger leader. Whistles or pyrotechnics must not be used. They will allow the enemy time to react.

- As soon as the enemy is hit, he reacts. The ambush force has only a few seconds to destroy the enemy in the kill zone. He will quickly recover from the initial shock and leave the kill zone--either with a direct counterattack or withdrawal, Subsequent fires and other banks of Claymore mines must be planned.
- The leader always executes the ambush, except when a member of the ambush knows he has been discovered. He then has the authority to execute--with killing fire, not by yelling.
- The cease-fire must be controlled by the ranger leader. A whistle or other device may be used to get attention, and then cease-fire is signaled.

(4) **Security.** The flanks and rear of an ambush site are open to counterattack. Flank and rear security may be enhanced by -

(a) Echeloning in depth.

(b) Designating sectors of observation.

(c) Positioning of RSTA devices.

(d) Enforcing noise and light discipline.

(e) Having a good withdrawal plan.

(f) Securing routes of withdrawal.

(g) Executing with speed and violence.

(h) Positioning security forces to sea] off the ambush area.

(i) Having good camouflage.

(5) **Simplicity.** A simple, direct plan improves the chance of success. The ambush plan must be clear yet concise to offer the greatest likelihood of success. For example:

(a) Mission statements for security, support, and assault elements must be clear, concise, and direct.

(b) Tasks to be performed by the ambush elements should be easy to understand.

(c) Contingency plans should be simple.

(d) Routes into positions and withdrawal routes should not cross. They should be the shortest, most secure routes.

(6) **Training and self-discipline.** Precise and violent execution is how ranger units win against enemy forces. All advantages must be exploited. Discipline must be strict. There must be no sleeping, talking, eating, or smoking in the ambush site. If an ambush is to be set up for long periods, then the elements of the ambush must be pulled back at set times to the ORP for rest. Extended ambushes of 24, 36, or 48 hours require six- or eight-hour shifts. It may take a company to man an extended platoon ambush position. Tired troops cannot man an ambush well. Other vigorous operations cannot be performed all day and have an alert ambush all night.

(7) **Maximum use of specialized equipment.** This includes the most sophisticated weapons, communications equipment, and RSTA devices. The effectiveness of any mission is enhanced by the use of this equipment.

(8) **METT-T.** Every ambush mission must be evaluated by the ranger commander IAW METT-T to design an ambush that works.

(a) **Mission.** Clearly state the mission--for example, stop enemy resupply, ambush enemy armor columns, or capture personnel or equipment.

(b) **Enemy.** State the type of enemy units--for example, guerrillas, main force, logistics carrying parties, or armor units. What is the size of the force, its habits when attacked, and type of support? What is the enemy night vision ability and expected time of movement?

(c) **Terrain.** Is it wooded, mountain, jungle, or urban? Some good ambushes can be set in a city or suburb. Are the routes available to the enemy suitable to his activity?

(d) **Troops.** Squads and platoons are the basic ranger ambush forces. An ambushing force should be able to engage a force two to three times its size--but evaluate enemy and terrain, A platoon ambushing an enemy tank company in daylight in open terrain is risky. But engaging that same column at night or in a forest may give the ranger platoon an advantage.

(e) **Time.** How long will it take to emplace the ambush? A complex, lethal ambush with plenty of "dirty tricks" takes hours to set up; a hasty ambush on a trail takes 15 to 30 minutes. How long is the ambush to be in place? What hours of the day, or at nighttime only? How soon must it be set up? Do not set up an ambush in darkness--even if it is to be executed in darkness. Try to allow every ranger to '6see" and "lay" in his weapon in daylight. Time is important.

c. **Organization of the ambush force.** A ranger unit conducting an ambush must be task-organized to perform the following functions: assault, security, and support. The ambush forces should be task-organized according to the TOE--by platoons, squads, and fire teams. The TOE should not be changed to create smaller elements for an ambush. The TOE formations may be reinforced with machine gun, recoilless rifle teams, or a 60-mm mortar squad.

(1) **Assault.** The elements assigned the assault mission either move directly into their positions or through a release point. The mission may include any combination of the following:

(a) Conduct the main assault.

(b) Halt an enemy's motorized column or any moving target,

(e) Kill or capture personnel.

(d) Recover supplies and equipment.

(e) Destroy vehicles and supplies.

Search teams are not always used. The ranger leader must decide how and when to use search teams. When rangers leave the security of their well-chosen, concealed ambush position, they are subject to the fires of the enemy who may also be hidden and ready. If it is at night, do not throw tripflares or shoot illumination to light the search area as this will illuminate and expose rangers to the enemy. Always assume there is hidden enemy--the ambush will not kill them all. Night observation devices (NODs) or a red filtered flashlight taped to M16s should be used to quickly search in ranger buddy teams. If the return fire from the enemy is great or if the ambush missed the main body, then the leader may choose to break contact and leave without searching the kill zone.

(2) **Security.** The elements assigned the mission of security may move to their positions

directly or by way of a release point. Their missions may include any or all of the following:

(a) Secure flanks, rear, or ORP.

(b) Provide early warning.

(c) Seal off the kill zone to prevent the enemy from escaping or reinforcing.

(d) Assist in executing the ambush.

(e) Cover withdrawal of main ambush force.

(3) **Support.** The units assigned a support mission provide fires that may include employment of--

- Heavy automatic weapon fires.
- Antitank fires.
- Mortar fires.
- Mines.
- Flame munitions.

d. **Ambush site.**

(1) **Choose an ambush site.** When choosing an ambush site, all sources of information must be used to enhance surprise, exploit the enemy's weak points, and take advantage of the terrain. An analysis of the terrain is also based on the other factors of METT-T to include maps, aerial photos, and, if possible, personal reconnaissance. Emphasis is on--

(a) Natural cover and concealment for the ambush force.

(b) Routes of entry and withdrawal (at least two) that should be direct and easy to reach.

(c) Good observation and fields of fire.

(d) Harmless-looking terrain.

(e) Few enemy escape routes.

(f) Limited enemy reinforcement ability.

(g) Nearby assembly or rendezvous area.

(h) Terrain that will canalize enemy into kill zones, and natural obstacles to keep him there.

NOTE: Try to select a site covered by friendly supporting indirect fires.

(2) **Take advantage of the terrain.** Emphasis must be on exploiting all natural cover and concealment afforded by the terrain. Site the ambush and individual positions based on the terrain rather than trying to adapt the terrain to a fixed geometric design.

(3) **Restrict enemy movement.** Restricting enemy movement by natural or man-made

obstacles should also be planned.

e. **Types of ambushes.**

(1) Ambushes have two categories for ranger purposes--area ambush and point ambush.

(a) Area ambushes may be set up by ranger platoons, companies, or battalions. They are used to interdict enemy movement in a given area or inflict casualties on his forces. Area ambushes consist of a series of point ambushes. The size and location of the ambushes are dictated by the METT-T analysis.

(b) Point ambushes are set at the most ideal location to inflict damage on the enemy. Such ambushes must be able to accept the enemy force from more than one direction. The ranger force levels may enable it to execute an ambush in two or three main directions. The other directions must be covered by security that gives early warning of enemy attack.

(2) The basic ambush is linear (Figure 6-1). An ambush is set on an expected avenue of approach. This ambush can accept contact from three basic directions--left, right, and front. The rear is secured by a security team.

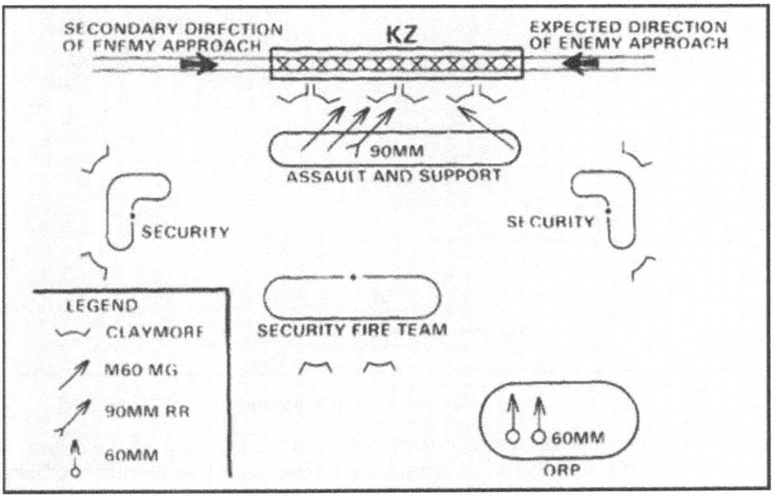

Figure 6-1. The linear ambush.

(3) A variation of the linear ambush is the L-shaped ambush (Figure 6-2). The L-shaped ambush is formed with the base (bottom) of the L perpendicular to the expected enemy direction of advance. This is a good ambush for a road, jungle trail, or an area where the enemy is canalized and his approach route is known. The L-shaped ambush can handle an enemy coming from the expected direction (toward the base of the L) from the front (stem

of the L). It is less effective against an enemy formation that comes from the opposite direction. The ambush must be executed when the enemy main body has cleared the base of the L. The flank security must protect the rear of the base of the L. In the L-shaped ambush, weapons must be sited to avoid direct or ricochet fire into friendly forces. The security elements must be considered as fire fans are being planned.

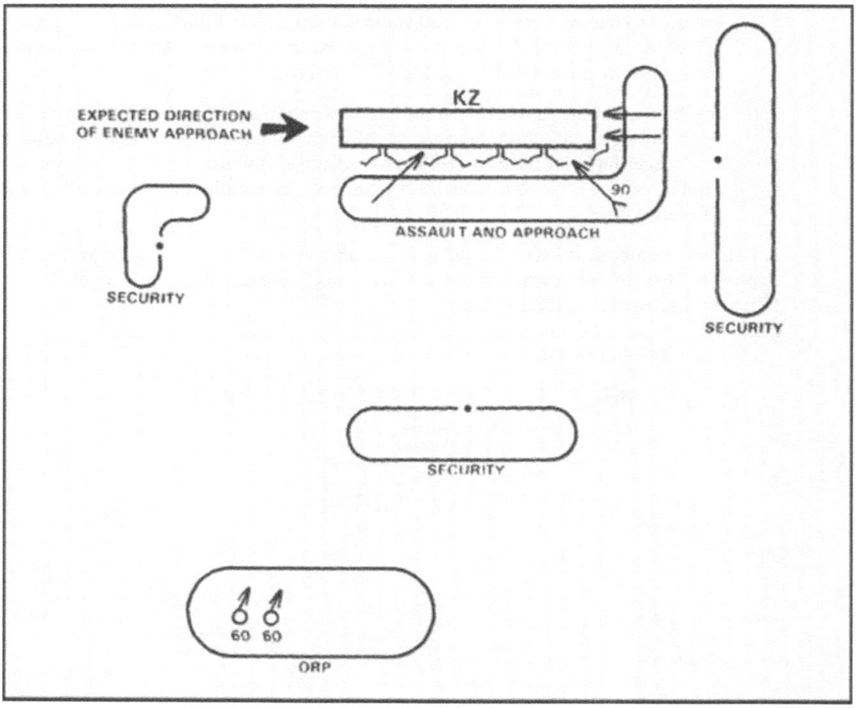

Figure 6-2. The L-shaped ambush.

(4) Other ambushes (the Z, the X, the V, and so on) are all variations of the linear and L formations (Figure 6-3). The more advanced ambush formations are difficult. Units must be expert in the basics before they are ready to employ such techniques.

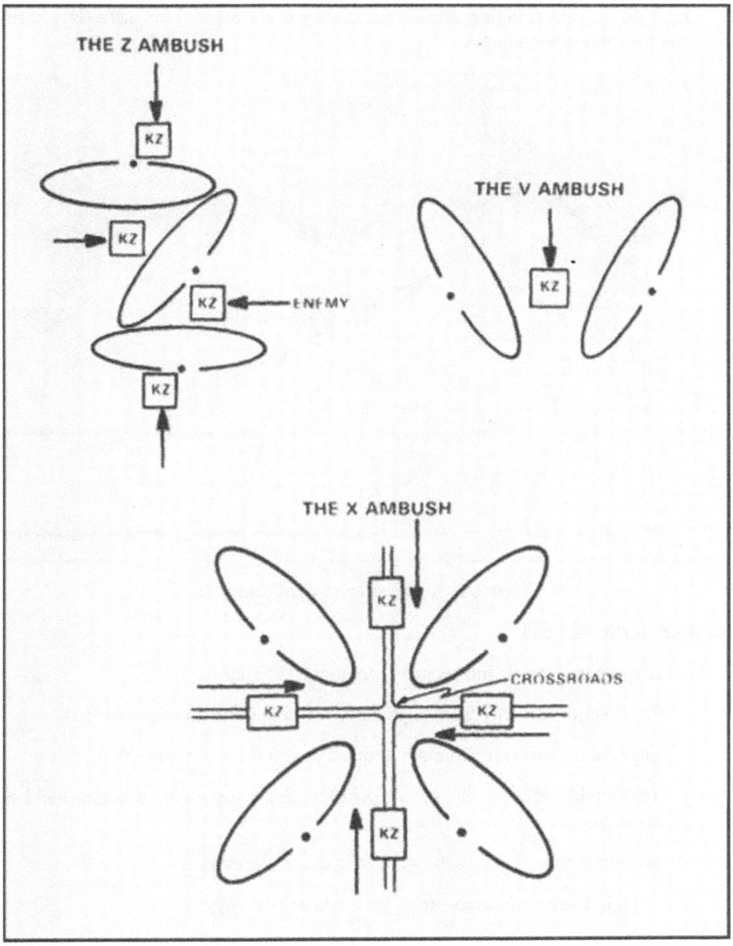

Figure 6-3. Other ambush variations.

(5) Another type of point ambush is the mechanical ambush (Figure 6-4). It consists of Claymore mines set in series with a double-ring main. It is command detonated or detonated by a triggering device activated by the enemy. Mechanical ambushes are normally manned. Rangers prepare to engage the enemy with direct fire if the mechanical ambush does not detonate or if it is wholly or partially ineffective. Mechanical ambushes are an effective way to interdict a large area using a small ranger force. If the mechanical ambush is effective and

rangers do not reveal their presence, the enemy is confused. This has a devastating effect on his morale and effectiveness.

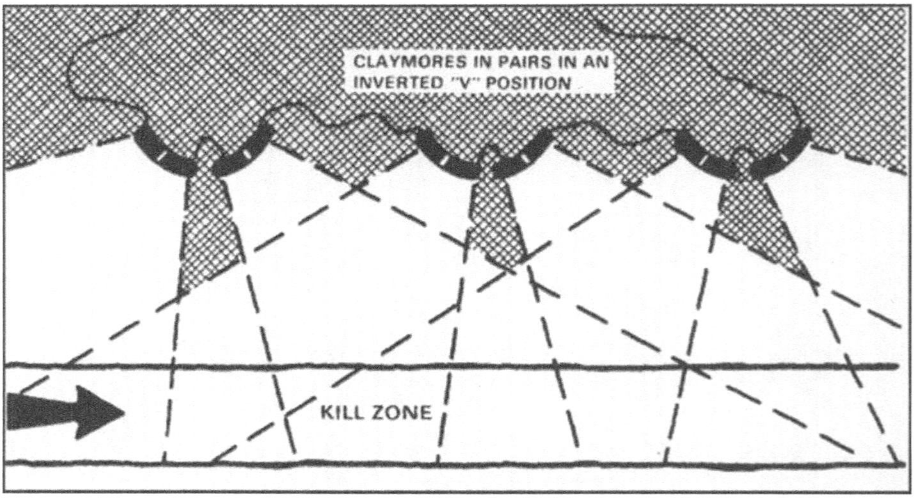

Figure 6-4. The mechanical ambush.

f. **Execution of the ambush.**

(1) **Occupation.** Stealth and security are important factors.

(a) Position security teams and early warning detection devices first.

(b) Use the rest route to main ambush position consistent with security.

(c) Quickly occupy the ambush position and set up communications and signaling devices.

(d) Position key weapons (automatic and antiarmor).

(e) Rig Claymore mines, tripflares, and booby traps.

(f) Ensure that all weapons are correctly positioned. Assign sectors of fire to provide mutual support and cover dead space.

(2) **Camouflage.** Each man must be hidden from the enemy, During preparation for the mission, each man camouflages himself and his equipment, and secures his equipment to prevent noise. At the ambush site, positions are prepared with minimum change in the natural appearance of the site. All debris resulting from preparation of positions is concealed to prevent evidence of occupation.

(3) **Movement.** Movement is kept to a minimum. The number of men moving at a time is closely controlled. Every man is as quiet as possible, especially at night. Light discipline is rigidly enforced at night. Smoking is forbidden.

(4) **Signals.** Audible and visual signals, such as whistles or pyrotechnics, must be changed often to avoid setting patterns and alerting the enemy. Three or four simple signals are needed to execute the ambush. Signals are used to--

(a) Provide early warning of an enemy approach. A signal by the security force to alert the patrol leader to the correct direction of enemy approach may be given. This includes arm-and-band signals, radio, or field telephone.

(b) Initiate the ambush. This may be the detonation of mines or explosives, Fire is then delivered at once in the heaviest, most accurate volume possible. Properly timed and delivered fires add to the achievement of surprise, as well as to the destruction of the target.

(c) Lift or shift fires, Signal for lift or shift fires, if the kill zone is to be assaulted. Voice commands, whistles, or pyrotechnics may be used. When the kill zone is assaulted, the lifting or shifting of fires must be as precise as when starting the ambush. Otherwise, the assault is delayed and the enemy has a chance to recover and react.

(d) Withdraw. A signal for withdrawal can be voice commands, whistles, or pyrotechnics.

(5) **Objective rally point.** The ORP is located far enough from the ambush site so that it will not be overrun if the enemy manages to attack the ambush unit(s). Existence loads are normally left in the ORP. Routes of withdrawal to the ORP are scouted (when possible by each man), Withdrawal routes should provide cover and concealment for the unit and hinder enemy pursuit. Withdrawal routes are a main consideration in the selection of the ambush site. They may be the key to survival after execution of the ambush. On signal, the ambush force quickly (but quietly) withdraws to the ORP. If pursued, the withdrawal may be by bounds, with mines or hasty ambushes used to delay pursuing forces.

(6) **Ambush variety.** If one method of ambush is used predominantly, the enemy will develop an effective defense against it and will be affected less by the shock of the ambush since he knows what to expect. More than one method of ambush should be used. No single method will fit all combinations of terrain, equipment, weather, and enemy capabilities. Variety should also be used when signaling - for example, use multiple signals, both audible and visual. Use weapons fire, mines, and RSTA when possible and vary signals to avoid compromise.

(7) **Swift action.** Speed in the execution of the ambush and the withdrawal should prevent enemy reaction forces from engaging the ambush force. Speed is often a shield against casualties and failure. When there is contact with reaction forces, speed may enhance quick disengagement. Ranger units must not only perform better than the enemy but also move faster than the enemy can react.

g. **Successful ambush.** Emphasize the following:

(1) Intelligence to ensure the enemy is ambushed at a time and place when he least expects or is least prepared to fight. This is preferred during reduced visibility to achieve surprise.

(2) Detailed planning and thorough training and rehearsing of all elements in all phases of the ambush. This ensures maximum shock effect through swift, precise execution.

(3) All available RSTA devices to permit daytime effectiveness at night when moving, shooting, or detecting enemy movement. To avoid detection, active RSTA devices should not be used until after the ambush has been triggered.

(4) All available firepower with emphasis on antiarmor, area and automatic weapons, and precision-guided munitions.

(5) Speed to achieve surprise and enhance security of the force.

(6) Cover, concealment, and overall protection afforded by the terrain when moving or when occupying ambush positions.

6-3. Antiarmor operations.

a. Antiarmor operations are not primary missions for ranger forces. They lack antiarmor firepower and have limited antiarmor ability. Ranger units can, however, perform limited antiarmor missions on close terrain by interdiction or ambush. The antiarmor weapons they do have are man-portable, but the ammunition is too heavy to carry many rounds cross-country. A ranger battalion has ten Dragon missile launchers and nine 90-mm recoilless rifles. It is also equipped with light antitank weapons (LAWs) and laser target designators (LTDs) that can guide Copperhead rounds.

b. Every ranger must be skilled with the limited antiarmor systems available. Proper selection, fortification, and camouflage of antiarmor ambush sites, route selections, withdrawal procedures, and initiative are important. Armor kill zones (AKZs) are normally set up after infiltrating the ranger force behind enemy lines under limited visibility.

c. A ranger antiarmor force is organized into four elements: command and control, support, security, and armor-killer teams (Dragons, LAWs, 90-mm recoilless rifles). These elements function the same as a raid force. The exception is that the assault force is the armor-killer team with the mission of destroying enemy armor, and it will probably not enter the AKZ. After the ambush occurs, the ranger force withdraws to its ORP. Indirect fires (if available) are then brought to bear on the AKZ.

d. Ranger units may conduct antiarmor ambushes and interdiction operation by use of laser-guided Copperhead rounds or air-delivered precision-guided munitions. Air-delivered family of scatterable mines (FASCAM) can be used to provide the ranger force security against attack by armored forces.

6-4. Military operations on urbanized terrain.

a. The ranger regiment normally tries to avoid the manpower-intensive, high-casualty battles associated with MOUT. However, due to the nature of strike operations and the types of targets selected, the regiment may conduct MOUT for limited periods. As in all ranger operations, the ranger force commander tries to surprise the enemy, attack him from unexpected directions, and destroy him with a violent, precise attack. The maze of cellars, rooms, sewer systems, rooftops, and alleys common to urban areas presents many opportunities for the ranger force.

b. Offensive operations by a ranger force exploit the centralized planning and decentralized execution aspects of ranger unit training. Individual action and initiative are always channeled to accomplish the commander's intent. The ranger unit's offensive tactics are based on quick action and boldness. Attacks by the ranger force are short and sharp. Execution is marked by swiftness, daring, timing, and surprise. The enemy's sleeping habits, eating habits, and relief times are watched. launched almost always at night or under smoke screens, attacks are often made without mortars or artillery, relying on surprise for success. Snipers and sharpshooters are used to shift the balance of battle.

c. Deliberate defensive operations are not good ranger unit missions. However, if a ranger unit must defend for a short period, the close terrain of urban areas allows it to function well against a stronger force. The buildings in a city break up advancing enemy formations and force them to move along streets. The ranger force executes an active defense anchored on strongpoints, such as a building or a group of buildings. It launches surprise counterattacks against the enemy's flank and rear. The aim of ranger defense is to give the enemy no respite and to make him feel threatened by a sudden, deadly attack. Ranger units should be withdrawn from defensive positions in urbanized areas as soon as possible. Prolonged MOUT and the associated casualties quickly degrade a ranger unit's combat effectiveness.

6-5. Perimeter defense.

The Perimeter defense may be conducted to control key terrain, gain time or economize forces.

a. **Planning.**

(1) **Warning order.** The order must be issued soon after mission receipt. It should contain the mission, time schedule (including rehearsal and inspection times, mission time, and place for OPORD), uniform and equipment, and special instructions.

(2) **Operation order.** The leader normally gives the OPORD on the ground to be defended. Locations and sectors of fire for squads and crew-served weapons should be designated. If time permits, leaders walk the terrain with subordinates as the OPORD is issued. The OPORD covers security, communications, signals, target reference points (TRPs), armor kill zones, and priority of engagement rules for antiarmor weapons. The OPORD also states the priority of work.

(a) Set up security and communications.

(b) Position crew-served weapons and personnel.

(c) Clear fields of fire.

(d) Prepare fighting positions.

b. **Organizing.**

(1) Squads are positioned with overlapping and interlocking sectors of fire and observation.

(2) The M60 machine guns are positioned either with an assigned final protective line (FPL) of fire or with a primary direction of fire (PDF). The PI)Fs are employed on covered avenues of approach such as gullies. The M60 FPLs should interlock across the unit front.

(3) The 90-mm recoilless rifles and Dragons are positioned with frontal cover to engage attacking armored vehicles from the flank.

(4) Riflemen are positioned to support and protect crew-served weapons. Riflemen are assigned FPLs that interlock at a point just beyond hand-grenade range.

(5) Dead space is covered with Claymore mines, antipersonnel mines, and M203 fires.

(6) Antitank mines, positioned to provide close-in protection from armored vehicles, are covered by LAW fires.

(7) Leaders plan indirect fire for long-range suppression of armored vehicles and short-range protection from dismounted assault. Mortar or artillery final protective fires (FPF) are placed on the most dangerous dismounted avenue of approach.

(8) Range cards are prepared for all crew-served weapons. Squad and platoon leaders prepare sector sketches and fire plans.

(9) The ranger force camouflages everything from ground and aerial observation.

(10) Leaders designate alternate and supplementary positions.

c. **Executing.** Leaders control the distribution and concentration of fires in their sectors. They constantly consider the overlap of fires with units to their left and right.

(1) Squads and crew-served weapons engage targets in sectors assigned by platoon leaders.

(2) Each weapon is used in its best role:

(a) Rifle, against enemy infantry within 300 meters.

(b) Machine gun, against enemy infantry within 1,000 meters.

(c) Grenade launcher, against infantry within 350 meters and armored vehicles within 200 meters.

(d) Light antitank weapon, against armored vehicles within 200 meters.

(e) Recoilless rifle, against armored vehicles within 400 meters.

(f) Dragon, against armored vehicles within 1,000 meters.

(3) Most dangerous targets are engaged first.

(4) Squads shift fires as needed or as directed by platoon leaders.

(5) The platoon masses fire on selected targets while maintaining coverage of the entire platoon sector.

(6) Platoon leaders control fires.

(7) After repelling an enemy assault, the platoon consolidates and reorganizes the defense.

6-6. Breakout from encirclement.

A breakout from encirclement is conducted when units, operating behind enemy lines, find themselves cut off from friendly forces and surrounded by superior forces. Given ranger unit missions, the chance of operating as an isolated force behind enemy lines is great. The breakout is characterized by determining enemy weak points, deception, massing of combat power, and a direct attack for a violent and timely breakout.

a. **Characteristics and fundamentals.** The breakout from encirclement involves the following characteristics and fundamentals:

(1) Set up a hasty defense. Adjust the perimeter as needed and make contact between subunits. Assign each a specific sector.

(2) Plan fire support. Plan FPFs on likely dismounted avenues of approach. Mass antiarmor weapons on the most dangerous mounted avenue of approach.

(3) Conduct reconnaissance. Identify enemy weak points. Locate enemy armored forces and the routes they must use. Locate enemy automatic weapons.

b. **Planning.**

(1) Organize all support personnel into maneuver elements--for example, company headquarters, mortars, and antitank. Organize the ranger force into a rupture force, reserve, support force, and a rear guard.

(2) Issue a fragmentary order. It should contain current local enemy situation and cover all pertinent areas of the five-paragraph field order.

(3) Select rally points.

c. **Execution.**

(1) Reduce the defense perimeter. This is done slowly until the unit is massed for the breakout attempt.

(2) Create a diversion. This is done either by fire from the rear guard force or by a limited attack in a direction other than that of the main breakout attack.

(3) Break through the defense with the rupture force. Hold the shoulders of the breakthrough.

(4) Move the reserve through to exploit the breakthrough.

(5) Move the support troops through with casualties. Have them collect casualties en route.

(6) Have the rear guard follow the support troops. It may have to fight a delaying action and use fire and movement to break contact with the enemy.

(7) Consolidate and reorganize following the breakout.

(8) Have the rupture force break contact with the enemy once the rear guard joins.

6-7. Linkup operations.

During operations behind enemy lines, ranger units may be required to link up with other friendly ground forces. Linkup may occur in airborne, air assault, amphibious, or shore-to-shore operations. It may occur during the breakout of an encircled force or during the convergence of separate forces. It may involve conventional or unconventional US forces, allied forces, or friendly guerrillas. Rangers may be required to link up with other rangers, partisans, or friendly ground forces that are conducting offensive operations.

a. **Planning.** Ranger commanders plan detailed linkup operations to avoid friendly forces firing on each other. Coordination and planning of communications, recognition signals, fires, command relationships and responsibilities, and control measures are essential. Coordination should be done before the start of the operation.

(1) Site. The linkup site should be easy to recognize, have cover and concealment, be located away from prominent terrain the enemy might use, be defensible, and provide multiple access and escape routes. An alternate site should be planned in case of compromise.

(2) **Communications and recognition signals.** Frequencies, call signs, codes, visual signals, and alternates for each of these should be planned before departure of friendly lines. Radios may be used to report the location of each unit to the site, and occupation and securing of the site. A system of mutual recognition must be devised to keep friendly patrols from firing on each other. This may include visual signals such as arm bands, panels, colored lights, or RSTA devices.

(3) **Fires.** The headquarters directing the linkup sets up fire control measures and priority of fires. The most commonly used control measure is the restrictive fire line (RFL), beyond which one force may not fire without coordination with another force. The RFL may be adjusted as two units draw closer together. Such an adjustment needs close coordination and should be planned before the operation. Successive phase lines between the two units can serve as on-order RFLS, provided that neither unit comes too close to the RFL before it is shifted.

(4) **Command relationships and responsibilities.** These are set up by the headquarters directing the linkup. Ideally, liaison personnel are exchanged before the operation. The key to success is liaison and guides.

b. **Execution.** Each member of the ranger force must understand the plan. Rehearsals are necessary in order to avoid confusion during the linkup. One uninformed ranger can start firing and cause enough confusion that friendly forces could be identified as enemy. Actions must be quick since units are open to enemy attack while they try to link up.

(1) **Moving units.** A linkup between moving units is hard. As the units move closer to each other, the chance of them engaging one another increases. Therefore, the linkup units must adjust their movements to each other and continuously coordinate on a designated secure radio net. If possible, one or both units should come to a short halt before linkup.

(2) **Stationary units.** Moving units must know the positions of the stationary units. They must keep stationary units advised of locations. Stationary units guide the moving units to the linkup point by radio. The stationary units must be ready to accept and position the moving units smoothly and quickly.

6-8. Passage of lines.

A passage of lines is an operation in which one unit moves through another unit that is stationary and disposed in a tactical formation on a FEBA. Ranger forces may conduct a passage of lines to get behind the enemy. Movement in forward unit areas must be controlled, coordinated, and kept to a minimum. This avoids conflict with friendly troops or the activation of RSTA devices. Ranger units treat the positions of forward units as danger areas. They must be assumed to be under enemy surveillance in all weather or visibility. The ranger force is open to attack during the passage. Personnel and units may be more concentrated than is desired. Fires of the stationary unit may be masked for a limited time. The passing ranger unit may not be in a good position to react to enemy contact. Detailed reconnaissance and coordination are crucial to ensure that the passage is conducted quickly and smoothly.

a. **Principles.**

(1) **Coordination.** The ranger force commander, or his representative, must coordinate departure and reentry of friendly front lines (FFL). This is done either directly with the units through which the rangers pass or through designated staff agencies. Items to be coordinated include fire support, passage lanes, control measures, guides, communications, intelligence, CSS, and actions on enemy contact.

(2) **Reconnaissance.** The ranger force commander should make a ground reconnaissance of the area through which the unit will pass. He should pick the initial rally point (IRP) at this time. The reconnaissance should include observing the area just forward of the friendly unit for routes, danger areas, and obstacles. During reconnaissance or coordination, rangers should wear the same uniform as the troops in the forward unit. This will avoid attracting the enemy's attention.

(3) **Security.** Maintain security to avoid contact with the enemy while departing or reentering. It is important that the ranger force make no enemy contact at this time. Control is hard if a fire fight starts just forward of an FFL.

b. **Departure of friendly front lines.**

(1) Designate an IRP.

(2) Maintain security. Use an appropriate movement formation when departing the FFL.

(3) Do not move within the friendly unit's area without a guide.

(4) Make final coordination with the friendly unit commander.

(5) Count the ranger force members as they depart.

(6) Make a security and listening halt.

c. **Departure of friendly units.**

(1) The ranger force arrives at the forward unit and is met by a guide. The guide leads the rangers to an assembly area chosen by the ranger commander during initial coordination with the forward unit commander.

(2) No one should move, either singly or as a unit, anywhere in the FFL without a guide.

(3) The ranger commander makes final coordination with the forward unit commander. He determines if any changes have taken place since initial coordination and learns of any recent enemy activity.

(4) Before leaving to coordinate, the ranger commander gives instructions (called a contingency plan) for what should be done while he is gone. These instructions state where he is going, who is going with him, how long he will be gone, what to do if he does not return, and actions to be taken if there is enemy contact.

(5) On the commander's return from final coordination, he may issue a fragmentary order (FRAGO) to cover any changes.

(6) Three common threats and techniques for countering them during departure from the FFL are:

(a) **Ambush and chance contact.**

- The ranger force commander must avoid enemy contact while departing.
- The ranger force goes to the friendly side of the departure point where a point team is dispatched to see that the area forward of the barriers is clear.
- The point element should check the area out to the first covered position or an area large enough to allow the unit room to maneuver if it is hit while departing.
- The distance varies with the size of the ranger force and the terrain.
- The point team notifies the ranger commander when the area is clear. The rest of the unit is brought forward.

(b) **Indirect fire.** If enemy indirect fire is falling, the ranger commander should not halt his unit after final coordination at the forward command post. Instead, he should move it quickly through the FFL. The point element should be far enough ahead of the main body to provide security from ambush or chance contact.

(c) **RSTA.** If the enemy has radar, sensors, or night vision devices, they may be offset by the following countermeasures:

- Use a well-hidden departure point, such as a reverse slope or a dense woods.

- Infiltrate the unit through the departure area and have the rangers rendezvous in a defilade.
- Pass through during rain, fog, snow, or darkness to help conceal the passage.
- Employ electronic countermeasures.

(7) Have a security and listening halt after the ranger force has moved out of sight and sound of the FFL. This is a short halt to see if enemy is in the area, and to accustom the unit to the sights and sounds of the battlefield. Halt in a position that covers the unit from chance friendly small-arms fire. If the security and listening halt must be in an exposed position, have each man lie flat on the ground.

(8) During movement, have frequent halts to see that the unit is not being followed and that no enemy is in the area.

d. **Reentry of friendly front lines.**

(1) Set up and occupy a reentry rally point (RRP).

(2) Maintain security at the RRP and at the reentry point.

(3) Use as few personnel as possible to locate the reentry point while the bulk of the unit remains in the RRP.

(4) Meet a guide at the reentry point. Normally, a forward-of FFL password is used since it may be overheard by enemy.

(5) Count the ranger force members to ensure no infiltrators follow the unit, especially during reduced visibility.

(6) Give the friendly unit commander a spot report, providing combat information of immediate tactical value.

e. **Reentry of friendly units.**

(1) The unit is moved into a rally point near the reentry point. This rally point should be on a prominent terrain feature where the ranger commander can pinpoint his location, with respect to the reentry point.

(2) The forward unit is alerted by radio that the ranger unit is ready to reenter. A code word is used for security and brevity. The code word must be acknowledged by the forward unit before the ranger force reconnoiters the reentry point. This shows that a guide has been sent to the reentry point and is waiting for the unit.

(3) If the ranger commander is certain of the reentry point location, he moves the entire unit directly to the reentry point. He has the point element coordinate the reentry.

(4) An effective method of locating the reentry point is through the use of a thermal imagery device. It detects the body beat of the reentry guide, even though he may be well camouflaged or his location masked by smoke. If ground surveillance radar is available, it can be used to vector the ranger unit to the reentry point.

(5) The reconnaissance party should never move parallel to friendly barriers or move around the wire. If the reentry point cannot be found by the initial reconnaissance or if the reconnaissance party finds only the barrier wire, the unit should notify higher headquarters and move to another rally point to wait until daylight. The unit should not stay in a rally point from which a radio transmission has been made.

(6) When the reentry point is found, the ranger commander can bring the unit forward. If the reentry point is easy to find, he can call by radio for the unit to come forward.

(7) The guide leads the unit through the barriers to the security position coordinated with the forward unit commander. The unit halts in the security position, remaining in movement formation. The ranger commander then gives a spot report to the forward unit commander. The ranger commander tells him anything of intelligence value or of immediate tactical use to the forward unit.

6-9. Reconnaissance platoon operations.

a. The reconnaissance platoon organic to the ranger regiment consists of a platoon leader, a senior reconnaissance sergeant, and three five-man reconnaissance teams that include a communications specialist in each team (see Appendix A). This platoon is responsible for short-duration reconnaissance of objectives for the ranger force commander. This reconnaissance is routinely in support of future strike or special light infantry operations. The teams can also be used by the ranger force commander to conduct deep target acquisition and designation, or to confirm or disprove information reported by other sources. Rarely will teams be used to attack enemy targets.

b. The reconnaissance teams of the ranger regiment do not replace (or perform the same reconnaissance tasks associated with) long-range reconnaissance patrol (LRRP) units or long-range surveillance units (LRSUs) that may be organic to the corps or division. These units are organized, trained, and equipped for long-term passive surveillance and reconnaissance in enemy territory. The ranger reconnaissance teams are trained and organized to support the immediate intelligence-gathering needs of the ranger force commander. They are not to act as a human intelligence (HUMINT) collection asset for other operational headquarters.

c. The reconnaissance platoon headquarters trains, equips, and controls the employment of the three reconnaissance teams. It operates the communications stations needed to receive reports from deployed teams. In the absence of the S2, it debriefs team members upon extraction or exfiltration. It can provide one or more teams to support each ranger battalion on independent operations. The reconnaissance teams can--

(1) Infiltrate the objective area by parachute (HALO, HAHO, or static lines), helicopter, fixed-wing aircraft, SCUBA, small boat, foot, or other means.

(2) Remain undetected in the vicinity of the objective area up to five days.

(3) Perform reconnaissance operations employing a full range of night observation devices, infrared detection devices, unattended sensors, and photographic equipment.

(4) Perform demolition target analysis.

(5) Operate small watercraft and inflatable boats.

(6) Emplace unattended ground sensors, omni-directional navigational beacons, hand-emplaced expendable jammers, and electronic target designation devices.

(7) Collect combat information to satisfy priority information requirements and mission-essential elements of information. Teams report that information by use of long-range, secure, burst-transmission communications equipment.

(8) Perform DZ selection, marking and reception duties.

(9) Report objective area weather conditions.

(10) Perform highly selective, limited attacks or ambushes when so tasked.

(11) Link up with the main body of the ranger force in the objective area, or escape and evade the enemy in order to return to friendly lines.

(12) Act as part of the AAT during airborne operations.

d. There are many limits of the reconnaissance teams.

(1) Mobility is normally restricted to foot movement in the area of operations.

(2) Use of radio and other active electronic or optical devices makes the teams open to enemy detection.

(3) All supplies and equipment needed by the team must be carried on the initial insertion, since resupply increases the probability of detection by the enemy.

(4) Conduct of insertion and extraction operations deep in enemy-held territory is difficult due to the need for secrecy, timeliness, security, and accuracy of location.

(5) Medical treatment of team casualties is limited to individual first aid. Casualty evacuation is difficult unless linkup with the main ranger force has been achieved.

e. The training of reconnaissance team members parallels that given to all members of ranger units with emphasis on infiltration, communications, and reconnaissance skills. Extensive training is also given in survival, evasion, resistance, and escape (SERE) and advanced first aid. Training emphasis is placed on actions at the reconnaissance site. The reconnaissance teams must conduct their mission and remain unseen. Every team member is trained to consider the need for information balanced by the likelihood of detection. Detection by the enemy compromises future operations.

6-10. Reconnaissance planning.

a. Detailed planning ensures that the reconnaissance mission is successful. It is also vital to the survival of the reconnaissance team. This planning is conducted by the ranger force planning group assisted by elements of the JTF staff. The supporting air or naval commander, navigator, and electronic warfare officer may aid the ranger force staff in the planning.

b. Upon receipt of the mission, the team selected begins an intense preparatory phase at the REMAB or the ISB. There the team receives its initial mission briefing. The items covered allow the team to complete its detailed planning. The reconnaissance platoon leader reviews the team plan. He ensures compatibility and integration into the ranger force commander's plan for employment of the main force. Staff specialists at ranger regiment and battalion--such as those concerned with intelligence, weather, or communications--provide help.

c. Planning for the use of reconnaissance teams is conducted within the framework of the ranger planning sequence for the entire mission. There are five planning considerations for the employment of reconnaissance teams:

(1) Operations security must be considered in all staff efforts to provide maximum protection for the reconnaissance teams. This includes intelligence, CE, logistics, administration, and maintenance. Operations security must be integrated throughout every ranger force operation. It consists of four main categories of security measures: signal security, physical security, information security, and deception. All are related and must be considered at the same time for each operation.

(2) Some missions may require rapid deployment to the operational area, thereby dictating the quickest method of insertion. However, others may depend on maintaining total secrecy, regardless of time.

(3) The transportation means selected for the delivery and recovery of the teams depends on the mission. This selection is also based on the abilities, limits, and availability of mission support platforms.

(4) The distance to and from the objective area must be considered, as well as the distance from the insertion point to the actual target. Reconnaissance teams can move great distances over rugged terrain. However, this movement does require time and limits the team's ability to carry bulky or sensitive equipment.

(5) Consideration must also be given to intelligence information available concerning the objective area which includes, but is not limited to--

(a) **Enemy situation.** The enemy threat--his abilities, disposition, security measures, detection, and air defense systems - affects the selection of the means for delivery or recovery.

(b) **Terrain.** Land formations must be considered in selecting the method of infiltration and exfiltration. Terrain affects the selection of altitudes, approach and exit routes, landing areas, drop zones, and beach landing sites. Air infiltration routes that provide terrain masking are desirable in static line parachute operations.

(c) **Weather.** Seasonal weather conditions affect team delivery and recovery. Factors to be considered are temperature, precipitation, visibility, clouds, and wind. If para-SCUBA techniques are used, high surface winds and their effect upon surf conditions or period of reduced visibility may preclude the use of parachutes, inflatable boats, or swimming as entry or recovery techniques. These same conditions favor land infiltration and exfiltration, The AWADS reduces the impact of visibility as a limiting factor for air infiltrations.

(d) **Astronomical conditions.** Periods of sunrise and sunset, moon phase, moonrise and moonset, and twilight must be considered.

d. A reconnaissance team mission must be specific and support the mission. Information concerning the target may not be gathered because the risk of detection by the enemy is so great. Detection may preclude the conduct of the planned operation. If more than one mission is assigned, priorities are set, All team missions must be coordinated to avoid other friendly forces operating near the target. Ranger forces conduct operations at such depth behind enemy lines that no other US or allied forces are normally present. There is a chance that a friendly indigenous guerrilla force may be operating near the objective, but ranger reconnaissance teams do not normally make contact with them.

e. The reconnaissance platoon leader joins with the intelligence and operations sections of the ranger force headquarters in the initial planning for reconnaissance team operations. The method of operation while moving, communication procedures, reporting, surveillance procedures, and actions on enemy contact should be included in the platoon's SOPS.

f. The ranger force intelligence officer normally prepares the reconnaissance plan IAW the guidance provided by the controlling headquarters. He is helped by the operations officer and the reconnaissance platoon leader. Selected team members and a member of the unit providing the insertion and extraction means are briefed on the mission early in the planning phase. They should participate in the detailed planning to follow. The essential details of a reconnaissance plan normally include--

(1) Areas to be observed and recommended areas or positions from which this can be done. These positions are determined in advance of the insertions of teams. They are based on a study of terrain, road and rail nets, enemy order of battle, delivery means, operations plans of the controlling headquarters, and the desires of the ranger force commander.

(2) Load plans and procedures. This includes both insertion and extraction. Alternate sites are chosen for possible use. Primary and alternate DZs are selected, if applicable.

(3) The flight plan. This includes insertion and extraction flight routes and alternate routes.

(4) The movement plan to and from a patrol base. This includes the times, routes,. and destinations when individuals or small groups are required to operate away from the patrol base.

(5) The fire support plan. This includes the location and times of planned air strikes in support of the main body, and all planned fires by the main assault force. The reconnaissance teams should know the frequencies and call signs for tactical air support.

(6) The timing and ground tactical plan for execution of the main force operation. This includes disposition of the main force, special mission teams from the ranger force, or any other special operations force elements.

(7) The communications plan. This includes frequencies, reporting schedule, emergency reporting procedures, and alternate communications plans. It also includes the ground-to-air

communications plan and the friendly recognition signal to be used during linkup with the main assault force or other friendly forces.

(8) Plans for use of technical specialists or special equipment.

(9) The plan for the treatment and evacuation of sick or wounded team members from the objective area.

(10) The E&E plan. This includes routes out of the objective area, passwords, and alternate extraction zones or methods far from the objective area.

(11) Mission-specific elements of information desired by the ranger force commander, and any special or on-order missions the team must conduct.

g. The team leader uses specific steps in planning, preparing, and executing missions. These procedures are comprehensive, yet adaptable to any situation. Three of the most important steps are:

(1) **Issuance of warning order.** This action gives the team maximum time to prepare for the mission. It should be issued as soon as the team leader has been notified of a mission and updated after he has made his tentative plan.

(2) **Issuance of OPORD.** The team leader conveys the information and instructions needed to accomplish the mission. He requires a briefback from each team member.

(3) **Rehearsals.**

h. The specific steps used when planning missions are as follows:

Step 1. Study the mission.

Step 2. Plan use of time.

Step 3. Issue initial warning order.

Step 4. Study terrain and situation.

Step 5. Make tentative plan.

Step 6. Organize.

Step 7. Update warning order.

Step 8. Coordinate (continuous throughout).

Step 9. Consult aerial photographs and map updates.

Step 10. Complete detailed plan. (To be combined with paragraphs 3, 4, and 5 of the operation order.)

Step 11. Issue operation order.

Step 12. Supervise (throughout), inspect, rehearse.

Step 13. Execute mission.

i. A workable E&E plan must be developed. Such a plan enhances the survival of the reconnaissance team members. It ensures that they are psychologically prepared to conduct the mission, by knowing there is a system and plan for returning them to friendly lines. The E&E plan must include a section describing the actions of the reconnaissance team members should their insertion craft be disabled en route to the objective area. It should also include what to do if the team is unable to link up with the main ranger force, or if the main ranger force has not been inserted into the objective area.

j. Rehearsals and briefbacks are best for finding flaws in procedures or planning. Thorough coordination of all procedures is essential. The exact type of aircraft or other means that will be used for insertion should be used for training. Rehearsals should be conducted under terrain, astronomical, hydrographical, and meteorological conditions close to those to be met on the operation. The team leader conducts a briefback with the platoon leader. This lets the platoon leader decide if the team is ready for the mission. It also allows him to make changes.

6-11. Execution of insertion.

a. Reconnaissance teams normally conduct blind insertions. Blind insertions are amphibious, heliborne, or parachute drops onto unmarked DZs without help from a DZ-marking party or a CCT. Blind insertions often depend on favorable light and weather conditions in the objective area. Favorable conditions for the reconnaissance team often means low-lying cloud cover, fog, and darkness. The teams are able to operate during reduced visibility by the extensive use of night observation devices. The team's extensive training and high level of land navigation skills allow it to quickly traverse rugged terrain and not be detected.

b. Team insertion into the objective area often needs a cover and deception plan. Team members can conduct HALO and HAHO operations. Teams can be inserted into areas not normally acceptable as a DZ for conventionally trained parachutists. By the use of HAHO techniques, reconnaissance teams can assemble in the air and glide to a landing at a location far from the drop point.

c. If reconnaissance teams need special or bulky equipment that cannot be carried on the individual team member, it can be airdropped, hidden near the DZ, and retrieved later. Such container drops are often used during low-level static line insertions. The high-altitude airdrop resupply system (HAARS) can be used to deliver such items at the same time as reconnaissance teams are inserted by HALO.

CHAPTER 7

Combat Support

Combat support is used to enhance the combat power of the maneuver companies. Combat support elements may be organic or nonorganic. They can conduct a variety of missions in providing indirect and direct fire support, security, intelligence, and maneuver and fire control.

7-1. Intelligence support.

a. The success or failure of a ranger mission often depends on the accuracy, detail, and timeliness of intelligence information. Both deliberate and quick-response missions need specific information that can be obtained only by a multidiscipline collection and analysis agency. An active interface with all levels of the intelligence system helps get the needed information quickly.

b. The theater or JTF commander is responsible for providing intelligence support. The 1st SOCOM also provides information and analyzed intelligence to the ranger force commander. The main source of processed intelligence is the Echelons Above Corps Intelligence Center (EACIC), similar to the division level military intelligence (MI) all-source production center (ASPC). The EACIC provides intelligence, security, and electronic warfare (EW) interfaces (see Figure 7-1). It is through the EACIC that ranger commanders can request support from national, joint, combined, or CONUS intelligence analysis systems.

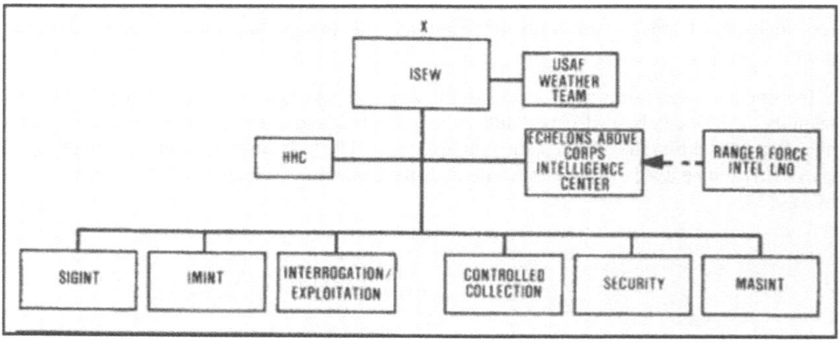

Figure 7-1. Intelligence at echelons above corps.

c. Intelligence requirements for ranger operations are unique, highly sensitive, and compartmented. The intelligence officer of the 1st SOCOM and the appropriate 1st SOCOM regional planning cell coordinate with theater intelligence agencies to provide needed information.

d. The ranger regimental staff and the ranger battalion staff have personnel in their intelligence sections to correlate and analyze information from many sources.

e. The reconnaissance platoon, organic to the ranger regiment, maintains surveillance of the objective area before the insertion of the main ranger force. It reports any last-minute developments that would affect the ground tactical plan. The platoon can also be tasked to determine any mission-specific items of essential information that the ranger force commander needs to complete his plan. This platoon is a highly specialized unit that is reserved for use by the ranger force commander. It is not a HUMINT collection agency for use by the theater or JTF commander. Reconnaissance teams normally link up with the ranger force in the objective area and are extracted with the main body. If this linkup cannot be made, they will exfiltrate enemy territory. They may use escape and evasion nets set up by other special operations forces or Department of Defense (DOD).

f. The ranger regiment normally places a liaison team at the headquarters of the command having OPCOM of a ranger force. This team provides command and control communications and liaison. It puts an intelligence liaison officer (LNO) at the EACIC or the appropriate ASPC (see Figure 7-2). This provides coordination to ensure that the intelligence needs of the ranger force are relayed to the intelligence processing center, and that the analysis is on the specific needs of the ranger force commander. This LNO coordinates with another located at the corps or EAC targeting center. Close coordination between the regiment, the intelligence analysis center, and the targeting center is essential to effective employment.

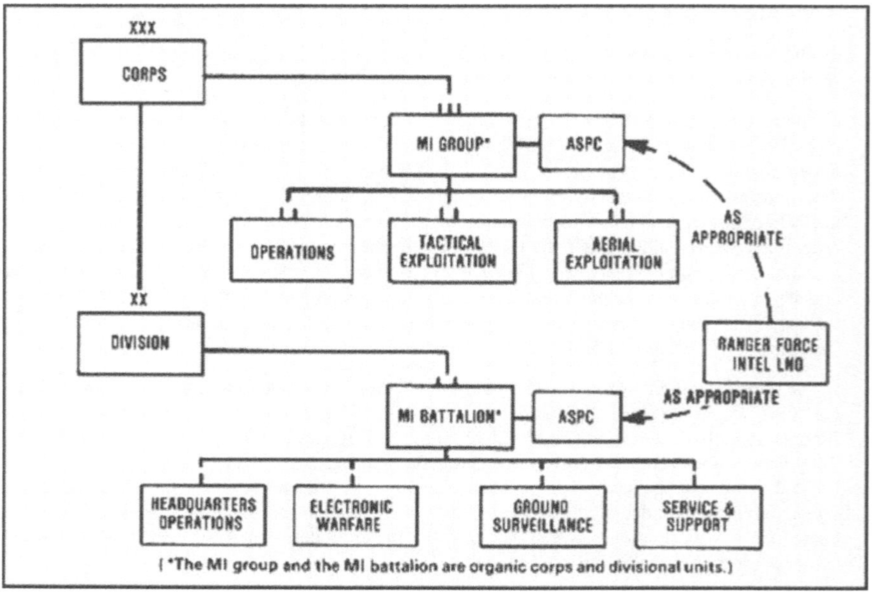

Figure 7-2. Intelligence source at corps and division.

7-2. Electronic warfare support.

a. Depending on the nature of the target and enemy abilities, a ranger mission may need support from EW units. This is mainly true during insertion. Electronic warfare operations disrupt or destroy the enemy's command and control of his forces and weapons systems, and retain friendly use of the electromagnetic spectrum. Electronic warfare is also used to support deception operations. They mislead the enemy by manipulation, distortion, or falsification of indicators, and persuade him to react in a manner prejudicial to his own interests.

b. Active jamming and chaff dispersal can prevent enemy early-warning radar from detecting the airborne force or determining its route. False transmissions can aid deception plans or feints used to help the ranger force's entry behind enemy lines. Selective jamming and imitative deception can disrupt the enemy command and control nets. This prevents the enemy from reacting to the presence of the ranger force in time to prevent mission accomplishment, or slows and interrupts his deployment of reaction forces.

c. Ranger elements, or other military forces, may conduct closely coordinated direct attacks against enemy EW sites during a ranger force's insertion or extraction. Not only does this reduce the enemy's ability to direct EW against the ranger force command, control, and communication system, but it misleads and confuses the enemy as to the true ranger objective.

d. The JTF or theater commander is responsible for planning the EW program to support a ranger mission. Planning and coordinating elements from the 1st SOCOM may also help. The EW program must be closely coordinated so as not to disrupt friendly air-to-ground or ground-to-ground communications, or to reveal the ranger force intentions.

e. The ranger regimental communications-electronics (CE) officer and those of the ranger battalions ensure that the required coordination concerning jammed frequencies, codes, authentication tables, and so on, is completed. The ranger regimental and battalion operations personnel ensure that the efforts of all EW units are combined to support the ground tactical plan.

7-3. Communications support.

a. Ranger operations are supported by secure long-range, lightweight, real-time, high-frequency (HF), and satellite communications. Effective long-range communications provide command and control links between deployed ranger units and the controlling headquarters. The JTF, ARFOR, or theater commander is responsible for communications between the controlling headquarters and the ranger regiment. The special operations force commander may also provide secure communications terminals to the ranger regiment or a deployed battalion.

b. Secure AM, FM, and SATCOM radios are the primary means of communication within the ranger regiment. Within the ranger battalion, AM and FM radios provide communications to company, platoon, and squad level.

c. During certain missions, specially trained and equipped quickreaction elements (QRE) deploy with the ranger force to provide secure communication links to the special operations force commander. These JCSE elements operate on either SATCOM or tactical satellite terminal (TACSAT) channels or through an airborne communications relay platform (see Figure 7-3). Depending on the mission, an airborne battlefield command and control center (ABCCC) aircraft or a joint airborne communications center/command post (JACC/CP) may be used. Their communications systems can operate at all levels of the national chain of command to permit a quick response to the tasking authority.

Figure 7-3. Airborne communications relay platform.

d. When only one ranger battalion is employed and the regimental headquarters is not the controlling headquarters, the ranger regiment provides a liaison cell to the controlling higher headquarters. This liaison cell includes a communications element from the regimental communications platoon. It can provide secure SATCOM, teletype, and facsimile transmission support to the ranger force in the objective area.

e. If two or more ranger battalions are employed, the ranger regimental headquarters deploys and acts as the command and control headquarters. The regimental communications platoon would then provide another communications link to the special operations force commander.

f. The ranger regimental CE officer ensures that the necessary communications links are set up and coordinated. The many communication means and channels available provide for effective control of a deployed ranger force. However, communication means must be closely coordinated at all levels of command to control the complex operations of a ranger mission. Planning and coordination with supporting aviation, transport, fire support, medical, and logistical elements before an operation are vital to efficient communications. The regimental CE officer must consider the communication systems linking the ranger force and other services. Air-to-ground and ship-to-shore communications are vital and must be set up early in ranger operations.

7-4. Fire support.

a. **Planning.**

(1) The ranger regiment has limited organic fire support assets. The six 60-mm mortars in each ranger battalion (two in each rifle company) provide the only organic indirect fire

available to the battalion commander. Each rifle company also has three 90-mm recoilless rifles and three Dragon missile systems that provide a small measure of organic direct fire support.

(2) The missions assigned to the ranger regiment and its subelements normally require fire support from organizations outside of the regiment. To plan, coordinate, and control these fires, the ranger regiment is organized with personnel dedicated to these tasks.

(3) The ranger regimental headquarters has a fire support element consisting of one fire support officer (FSO), two noncommissioned officers, and two fire support specialists. They prepare the fire support annex to the regiment's operation order. They ensure that all available fire support is planned to best support the regimental commander's ground tactical plan. They act as the primary coordinator for all external fire support, such as artillery, armed helicopter, close air support, naval gunfire, and short-range air defense. They also coordinate with each ranger battalion FSO to ensure unity of effort. The fire support element coordinates the efforts of air force, naval, and marine liaison officers when they are attached.

(4) The regimental commander may assign part of the fire support element to a committed ranger battalion as an attachment to help in the integration and planning of fires. The ranger battalions request through the fire support element that more or different fire support means be integrated into the ground tactical plan.

(5) The ranger battalion headquarters has a fire support team headquarters consisting of the battalion's FSO, two noncommissioned officers, and two fire support specialists. The FSO is the prime fire support coordinator for the ranger battalion commander. He maintains coordination with the regimental FSO and any attached fire support coordination elements, such as the USAF tactical air control party or naval shore fire control party. The FSO performs the following:

(a) Advises the ranger battalion commander on all fire support matters.

(b) Recommends allocation of fire support.

(c) Prepares fire support plans.

(d) Assigns target numbers.

(e) Processes target lists from the ranger company fire support teams to eliminate duplication.

(f) Monitors and functions as net control on the ranger battalion fire support coordination (FSCOORD) net.

(g) Reports changes in the status of fire support units to the ranger battalion commander, staff, and FISTS.

(6) When the ranger battalion is operating a single TOC, the FSO and his element are normally located there. When the ranger battalion is operating two TOCS, one TOC is normally manned by the FSO, the fire support sergeant, and a fire support specialist. The

senior fire support sergeant and the other fire support specialist would then monitor the FSCOORD net at the other TOC. The FSO ensures that any attached fire support elements also divide when needed to operate in a two-TOC configuration.

(7) The ranger battalion fire support team headquarters has three fire support teams assigned to it. These are normally allocated one to each of the ranger rifle companies. The company FSO and his team--

(a) Locate targets and request and adjust surface-to-surface fire support (mortar, field artillery, and naval gunfire).

(b) Plan fires to support the company ground tactical plan and prepare target lists.

(c) Coordinate fire support requests through the FSO for surface-to-surface and air-to-surface fires.

(d) Report battlefield information.

(e) Control air-to-surface fires of helicopters and prepare to assume control of fires from fixed-wing aircraft.

(f) Coordinate airspace use within the ranger rifle company operational area.

(g) Advise the ranger company commander on all fire support capabilities, limitations, and methods of employment.

(h) Inform all fire support units of target priority changes.

(i) Designate targets for laser-guided munitions.

(j) Assume operational control of, and administrative responsibility for, all augmentation fire support personnel (TACP, shore fire control party, and attack helicopter liaison teams).

(8) The company FSO is normally with the ranger rifle company commander. The FOs and their RATELOs are normally with the ranger rifle platoon leaders.

b. **Coordination.**

(1) Coordination of fires is a continuous process. It involves implementing the fire support plans of the ranger force commander and reacting to changing tactical situations. The success of a mission often hinges on how well fires are coordinated during the operation.

(2) The fire support planners within the many echelons of the ranger force become fire support coordinators during the actual execution of the operation. They use standard fire control coordination measures.

c. **Organic fire support.**

(1) **Indirect fires.**

(a) The 60-mortars constitute the only organic indirect fire support available within the ranger regiment. The mortar squads in each ranger rifle company normally operate as a section under the control of the mortar section sergeant. However, they can operate alone for short periods.

(b) The 60-mm mortar is used to destroy enemy equipment and light vehicles and to attack enemy personnel. Its main advantage is the ease with which it can be brought into operation against the enemy. This ease of employment allows the ranger force commander to concentrate combat power at the critical time and place. This helps to surprise and destroy the enemy.

(c) Mortar fires are employed by the ranger rifle company commander and coordinated by the company FSO. They are normally requested and adjusted by the FIST observers. The mortars are normally emplaced in a central position, and their fires are controlled by the fire direction center (FDC) using the indirect-fire method. The mortars may also use the direct-alignment or the direct-lay method.

(2) **Direct fires.**

(a) Each ranger battalion is authorized 90-mm recoilless rifles and Dragon missile launchers. These weapons constitute the heaviest organic direct-fire ability in the regiment. The antitank section in each ranger rifle company may operate as a complete section, but normally individual weapons are assigned throughout the company sector wherever their fires can best support the mission. These weapons can destroy most enemy armored vehicles and can penetrate the flank and rear armor of all enemy main battle tanks. The 90-mm recoilless rifle has a multiple projectile round that is effective against exposed enemy personnel. The HE round is highly effective against buildings and bunkers.

(b) The main disadvantage of direct-fire antiarmor weapons is their weight. These weapons, and their ammunition, are man-portable but place a strain on the unit when moving across rugged terrain.

(c) The ranger battalion is trained in the use of light antiarmor weapons, such as the US M72A2 LAW and the Warsaw Pact rocket-propelled grenade (RPG). The following provide protection for the ranger force against attack by enemy armored forces:

- The employment of light antiarmor weapons at all organizational levels.
- The placement of the 90-mm recoilless rifles or guided-missile systems on the most likely avenues of enemy-armor approach.
- An aggressive, offensively oriented employment.
- The ability to move into and across terrain impassable to armored vehicles.

d. **Field artillery fire support.**

(1) Many ranger unit operations are out of the range of supporting field artillery fires. Whenever such fires are available, they are planned for and integrated into the ranger force ground tactical plan.

(2) Coordination and exchange of call signs, frequencies, and target lists occur before insertion of the ranger force. Unless required by the tactical situation, ranger units do not normally receive augmentation forward observers from supporting field artillery units. The degree of mental and physical training necessary to execute certain ranger techniques requires more preparation than field artillery units are able to provide. The FIST from the ranger battalion requests and adjusts field artillery fires for the ranger force.

(3) Field artillery fires can be used to support the ranger force even if the objective area is out of range. Field artillery cannon fire and multiple rocket launch system (MRLS) fire can be used to suppress enemy air defenses. This helps the ranger force as it crosses the FEBA during airborne or air assault operations.

(4) Field artillery fires can be used to support the exflltration of ranger elements as they approach friendly lines. These fires can contribute to the deception plan and add combat power to feints used to support ranger operations.

(5) When in range of the objective area, field artillery units can be used to emplace field artillery delivered FASCAM to enhance the security of the ranger force.

(6) Copperhead rounds fired by 155-mm field artillery units can be terminally guided by the ranger FIST forward observers (FOs). They can attack hardened point targets or enemy armored vehicles by using a man-portable laser target designator (see Figure 7-4).

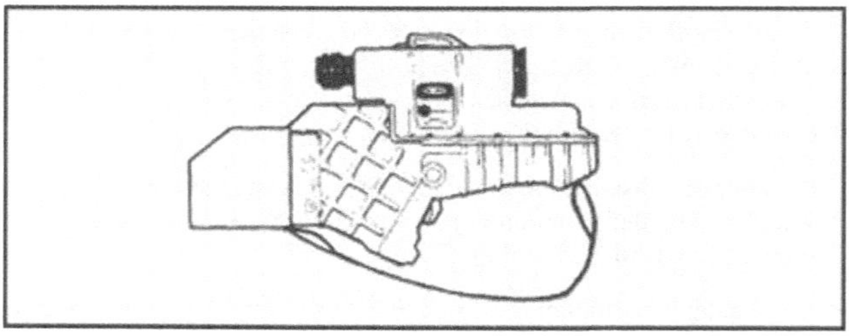

Figure 7-4. AN/PAQ-1 laser target designator.

e. **Aerial fires.**

(1) Aerial fire support is usually the prime means supporting the ranger force due to the distance behind enemy lines at which most ranger operations take place. Aerial fire support can be provided by either fixed-wing aircraft or helicopters.

(a) Fixed-wing aerial fire support may come from USAF, USN, or USMC units. The type of unit providing support, the aircraft, and the mix of ordnance carried all affect the fire support planning and coordination process. Some aircraft have a night and all-weather strike ability enabling them to support the ranger force during any level of visibility. Operations during poor weather that limits visibility to less than 3 nautical miles are still somewhat restricted. The ranger fire support coordinators must ensure that the correct aircraft are requested and employed effectively on the enemy. The

TACP directs and adjusts aerial fires in the objective area.

(b) The ranger force can use ground laser target designators to pinpoint targets for air strikes, as well as electronic navigation aids to permit nonvisual air strikes (beacon bombing). The ranger rifle company FIST or the TACP can control a laser-designated standoff air strike (see Figure 7-5).

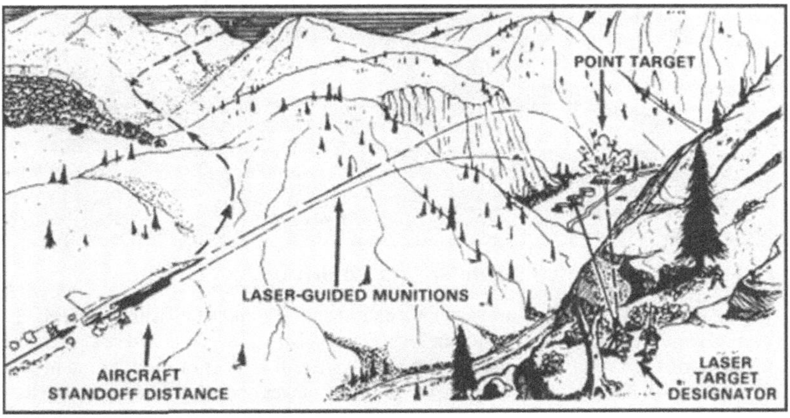

Figure 7-5. Standoff air strike.

(c) If the enemy ADA ability is not great or it can be degraded to a low level, the ranger force uses specially equipped and armed AC-130 aircraft (see Figure 7-6) for fire support. These aircraft provide an invaluable combination of firepower, night observation and illumination, communications, and long loiter time. A well-planned and executed suppression of enemy air defense (SEAD) program, coupled with ECM directed against enemy ADA units, normally permits the use of AC-130 aircraft.

Figure 7-6. AC-130 aircraft.

(2) The attack helicopter, armed with a mix of antitank guided missiles (ATGM), 2.75-inch rockets, 20-mm cannon, and 40mm grenade launchers, is accurate and responsive aerial fire support. The ranger force commander plans to use all sources of aerial fire to help accomplish his mission. However, the nature of ranger operations may preclude the extensive use of armed helicopters due to their limitations. Some attack helicopters are limited in range and lack an all-weather ability. They are restricted during operations at night by a lack of sophisticated night vision devices. They may lack the speed to go with the ranger force on some heliborne insertions. Attack helicopters may be used to escort and assist the ranger force as it crosses the FEBA. They may also be used to conduct feints and demonstrations to cover the insertion of the ranger force. As the MV-22 Osprey and the AH-64 Apache enter service, more use will be made of these aircraft to support ranger operations beyond the FERA, (See Figures 7-7 and 7-8).

Figure 7-7. MV-22 Osprey.

Figure 7-8. AH-64 Apache attack helicopter armed with Hellfire missiles.

(a) If attack helicopters are used to support a ranger operation, planned indirect fires are normally delivered along entry and exit corridors. Attack helicopters approach and depart the objective area using nap-of-the-earth flight profiles.

(b) Fires from armed helicopters are normally requested and controlled by the company FSO or one of his FOs, operating on a special ground-to-air net. The laser target designator may be used to precisely identify targets for the AH-64 Apache. Friendly unit locations may be marked by smoke, panels, lights, mirrors, or infrared sources.

f. **Naval gunfire.**

(1) During amphibious assaults or operations near a coastline, the ranger force may receive indirect fire support from naval gunfire. Normally, a ranger battalion would be supported by either a destroyer or a cruiser in a direct support role. However, the type and importance of the mission, the type of targets, the ships available, the hydrographic conditions, and the enemy capability determine how many and which type of ships are provided to support the ranger force.

(2) Naval gunfire is characterized by large volumes of highly destructive, flat-trajectory fire. Planned strikes in support of the ranger force may also include surface-to-surface missile fire. Some naval guns can fire a laser-guided projectile much like the Copperhead. This projectile, called a Deadeye round, is terminally guided the same as the Copperhead. Deadeyes can be guided either by an attached spotting party from the ANGLICO or by ranger LTD teams.

(3) When ranger units are being supported by naval gunfire, a shore fire control party is attached to the ranger force, This party normally consists of a liaison team and several spotter teams (forward observers). The liaison team is integrated into the operations of the fire support element (FSE) at the ranger battalion TOC. The spotter teams are attached to the ranger rifle companies. The shore fire control party LNO is the ship's representative to the ranger force commander, through the FSO. The spotter teams request and adjust fires from surface vessels. They can also request and control air strikes by carrier-based aircraft.

(4) The LNO and the spotter teams operate in the ground spot net, communicating with the ship by high frequency (HF) radio to request and adjust naval gunfire. The spotter teams communicate with the LNO using very high frequency (VHF) radios. The LNO also can communicate with aircraft using ultrahigh frequency (UHF) radios.

(5) Coordination and control measures that apply to naval gunfire are the same as for field artillery, with the addition of two terms, which are:

(a) Fire support area (FSA)--a sea area within which a ship may position or cruise while firing in support. It is labeled as FSA and numbered by a Roman numeral - for example, FSA VII.

(b) Fire support station (FSS)--a specified position at sea from which a ship must fire. This is very restrictive positioning guidance. It is labeled as FSS with a Roman numeral--for example, FSS VII.

g. **Air defense artillery.**

(1) Army air defense artillery units are rarely used to directly support the ranger force since

ranger operations normally take place deep behind enemy lines. However, it is possible that during the insertion or extraction phase of an operation, ADA units may support the ranger force as it crosses the FEBA.

(2) The ranger regiment's fire coordination element and the FSO in the ranger battalion are mainly concerned with adequate coordination of ADA fires. This prevents friendly units from being engaged. It allows the integration of ADA fires into the scheme of maneuver. The fire support planners ensure that enough coordination has been made so that ADA units are aware of the ranger force and know enough about its flight paths, routes, and altitudes to prevent engagement by friendly units. When conducting this coordination, care must be taken to pass only needed information to the ADA unit. To aid coordination, an ADA officer may be attached to the ranger force.

(3) The plan for a heliborne or airborne insertion and extraction of the ranger force must include the correct identification-friend-or-foe (IFF) codes, the flight profiles, the times of FEBA crossings, and the estimated number of aircraft making the crossing. There must also be a plan for crossing by straggler aircraft and battle-damaged aircraft that cannot reply to IFF interrogation. If possible, the routes selected should avoid concentrations of friendly ADA units.

(4) During an aerial crossing of the FEBA by a ranger force, the regimental headquarters may assign a liaison officer to the headquarters controlling the ADA fires in the crossing area. The LNO is then responsible for close and continual coordination. This ensures that the crossing is made without friendly ADA units engaging the ranger force while it is exiting or entering friendly lines.

(5) During ranger operations, the primary source of air defense fires is shoulder-fired antiaircraft missiles carried by specially trained individual rangers. These weapons provide the ranger force commander a limited defense against attack by enemy aircraft. The decision to take these weapons into the objective area depends on the factors of METT-T. Ranger operations should be completed before the enemy can react with ground forces. However, enemy aircraft may be able to react in time to threaten the ranger force. These lightweight missiles offer a highly effective, short-range counter to enemy air attack.

(6) The ranger force commander must define the weapons control status and rules of engagement for all short-range air defense (SHORAD) teams under his control. Priorities of target engagement must also be determined to prevent multiple engagements of a single enemy aircraft. The ranger battalion FSO and the attached ADA officer are responsible for the coordination of ADA fires within the objective area. Standard ADA weapons-control status and rules of engagement may be modified by the ranger force command.

7-5. Aviation support.

a. **Planning.**

(1) The ranger regiment may need extensive aviation support on some missions. The ranger battalion is organized and trained to move quickly to the objective area by air and to be resupplied by airdrop for short periods.

(2) Often the decision to execute a mission depends on the amount and type of available insertion and extraction aviation support. Normally, the theater commander or JTF commander allocates the type and amount of aviation assets to support a ranger operation. The theater or JTF commander's staff is responsible for the planning and coordination concerning aviation support. The ranger force commander must play an active part to ensure mission success.

(3) Due to the extensive training requirements and specialized operational techniques needed to support special operations, special air operations units in the Army and the USAF have been equipped and trained for the mission. These forces train with the ranger regiment and jointly develop tactics, techniques, and procedures of employment. These special units operate diverse aircraft that have unique capabilities. These help the ranger force commander to effectively employ his units. Command of these special units is maintained at a very high level in the chain of command. They are assigned to support a specific operation. Then they return to their parent unit's control.

b. **Coordination.** In addition to the staff of the theater commander or JTF commander, the coordination of aviation support occurs at all levels in the ranger force. These operations often are conducted in phases. A separate air mission commander is then designated. The AMC is responsible for the completion of each phase of the aviation mission, as well as the close and continuous coordination with the ranger force, commander. The ranger regimental air operations officer and the movements control officer are responsible for the planning, integrating, and controlling of aviation support.

c. **US Air Force.**

(1) The USAF provides special aviation support to the ranger regiment and other elements of the special operations force. The USAF has elements that are equipped with MC-130 Combat Talon aircraft (see Figure 7-9) and AC-130 Spectre gunships, as well as HH-53 Pave Low, UH-60, and UH-1 helicopters. These aircraft are specially equipped to conduct long-range infiltration, exfiltration, attack, and search-and rescue operations. The USAF also provides C-141B and C-130 aircraft for strategic and tactical airlift of the ranger force. There are specially trained USAF crews that can operate using AWADS and SKE. This gives them an excellent means of inserting the ranger force during reduced visibility.

Figure 7-9. MC-130 Combat Talon aircraft.

(2) The AFFOR commander provides the airlift control element (ALCE) support at the departure airfield. The ranger air operations officer ensures that coordination of aircraft load limits, air movement times, specific load plans, and manifests is conducted in an effective and timely manner. The air operations officer and the movements control officer ensure that a departure airfield control officer (DACO) from the deploying unit is designated to provide face-to-face coordination at the departure airfield.

(3) The USAF normally provides an airborne battle command and control craft. This is a specially equipped C-130-type aircraft that provides communications and data system down-links to ground stations.

(4) The terminal guidance needed for assault landings and air traffic control in the objective area is provided by a specially trained and equipped CCT. This element has habitual training and operations relationships with the ranger units. The USAF is also responsible for the specially trained and equipped weather teams that can be sent into the objective area with the CCT.

(5) The USAF Tactical Air Command (TAC) includes fighter, interceptor, ground attack, electronic countermeasure, early warning, reconnaissance, and airborne command and control squadrons, any of which may be used to support the ranger force. The Tactical Air Command provides the results of aerial reconnaissance to the AFFOR, who then processes it for use by the theater or JTF commander's staff. This reconnaissance may be by low-level photo, imaging infrared, side-looking airborne radar (SLAR), low-light-level television, or overhead photography.

(6) The Military Airlift Command (MAC) equips and trains special long-range SAR aircraft and crews. These aircraft, mainly helicopters, are equipped with sophisticated navigation, observation, and communications equipment, as well as extended-range fuel tanks and in-flight refueling capabilities (see Figure 7-10). The ranger regiment uses these aircraft for certain missions.

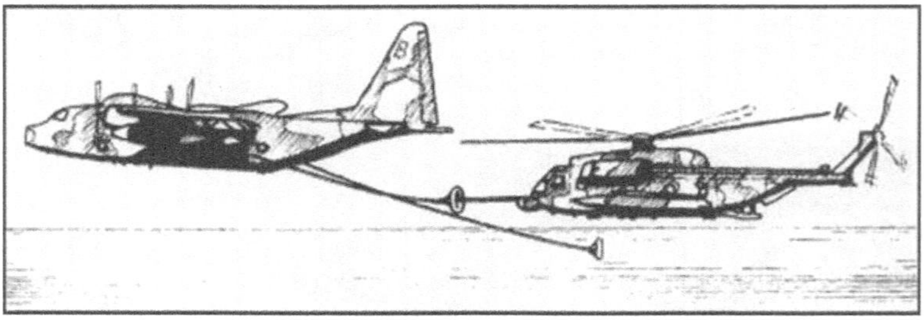

Figure 7-10. In-flight refueling.

d. **US Navy and US Marine Corps.** Both the USN and the USMC have units equipped and trained to support ground forces. The type of aviation support normally provided to the ranger regiment is close air support against targets in the objective area, or as suppression missions against enemy air defense installations. Long-range helicopters from the USMC may be used to support an amphibious operation or to insert special teams during ranger operations. The USN also has a search-and-rescue ability that may be used to support the ranger regiment.

e. **Special operations aviation units.**

(1) The US Army has specially equipped and trained SOA units tasked to support the ranger regiment, as well as other elements of the special operations force. These SOA units have a habitual training association with the ranger regiment and use dedicated aviation assets to -

(a) Insert, extract, and resupply the special operations force.

(b) Conduct armed escort, reconnaissance, surveillance, and electronic warfare in support of the special operations force missions.

(c) Provide airborne command, control, and communications for special operations force elements.

(d) Provide general support aviation during peacetime and contingency operations.

(2) Penetration of enemy airspace is normally done during limited visibility at very low altitudes. Penetrations are planned to maximize the use of existing terrain and exploit blind spots in enemy air defenses. On-board navigational aids are crucial to mission success. There normally is not any terminal guidance to the selected LZ or DZ. Deception operations of a tactical or operational nature help SOA units penetrate to the objective and reenter friendly lines. Advanced helicopters and short-takeoff-and-landing aircraft provide the help needed to do these missions.

(3) To survive enemy air defense systems and prevent mission compromise, SOA units avoid detection. Detected aircraft may reveal the location of supported special operations force ground forces or may be destroyed by enemy ADA fires. The SOA normally avoids detection by operating at night without illumination. This condition requires flight using sophisticated night vision devices. The SOA also operates in other conditions of reduced

visibility, such as clouds, fog, rain, snow, smoke, and dust storms.

(4) Before an operation, the ranger force commander may need to place reconnaissance teams near the objective. Army SOA assets can be used to transport ranger reconnaissance teams.

(5) Elements of SOA provide the ranger regiment with short and intermediate range penetration, resupply, and extraction capabilities. Long-range aircraft must be provided from USAF special operations assets.

(6) The ranger regimental or battalion commander is normally provided a liaison aircraft while the ranger force is in the REMAB. This ensures that he can travel quickly to the dispersed headquarters of the agencies providing support for the mission. This aircraft and crew are provided by the theater or JTF commander from EAC general support aviation assets. It may be a helicopter or fixed-wing aircraft, depending on the location of the REMAB to a suitable landing field.

(7) Army helicopters and fixed-wing aircraft may be used to provide SEMA aerial reconnaissance; radar, infrared and photographic imagery; and electronic intelligence support to the ranger force commander. This support is normally provided by the aerial exploitation battalion of the theater Army military intelligence group.

7-6. Engineer support.

a. The ranger regimental and battalion operations officers plan and request engineer support for ranger operations.

b. The most common engineer-type tasks needed during a ranger operation are the destruction of enemy facilities, equipment, and material by use of conventional explosives; and the interdiction of major lines of communication. To reduce the need for external support, the ranger regiment trains selected individuals in advanced demolition techniques. All members of the ranger regiment are well trained in the basic employment of conventional demolitions. Therefore, there is no need to attach combat engineers to conduct normal demolition missions. If specialized skill or knowledge is needed for a specific mission, qualified personnel can be attached.

c. During the occupation of the REMAB and the rehearsal for an operation, the ranger regiment needs construction engineer support to build simulated target areas. The simulated target areas should be camouflaged to prevent the enemy from recognizing the target. The construction engineers augment the ranger unit's self-help construction efforts. Limited-facilities construction and improvement may also be needed at the REMAB to provide security and isolation. The theater commander or JTF commander provides the construction engineer support.

d. Certain missions, such as those of interdiction, may need engineer special weapons teams. The teams are provided by the theater commander or JTF commander, or they may come from other special operations force assets. They are attached to the ranger force for a specific mission and released at the completion of the extraction and debriefing phase.

e. Some operations may need the ranger force to use captured enemy heavy construction equipment. Selected members of the ranger regiment are trained to operate such equipment.

However, construction engineer personnel may be attached to the ranger force for this mission.

CHAPTER 8

Combat Service Support

Combat service support consists of the logistical and administrative effort to maintain the battalion's ability to fight. The commander forecasts his needs for CSS. He employs CSS units to ensure accomplishment of the mission.

8-1. Ranger support element.

a. The ranger support element is a non-TOE organization formed from elements of CSS units collocated with CONUS-based ranger units augmented by selected personnel from the host installation. The mission of the RSE is to provide all the CSS needed to outload and deploy a ranger unit from CONUS. The personnel designated as part of the RSE have the same alert notification, reaction, and security requirements as members of the ranger unit they support. The RSE is normally built around a command and control group provided by a major CSS unit on the ranger unit's host installation. The remaining CSS elements and teams are then designated. An RSE consists of--

(1) An emergency operations center liaison group that provides timely coordination between the ranger unit and the functional elements of the RSE.

(2) A command and control group that coordinates and directs the actions of the functional elements.

(3) A supply section responsible for the timely issue of all Class II, III, IV, and VII items. This section maintains a stockage of contingency items marked for use by the resident ranger unit. It also issues items out of common stocks.

(4) A transportation section moves equipment and personnel from the unit area to the departure airfield and picks up equipment from local storage sites. It may also operate material-handling equipment and trailers to support loading of aircraft.

(5) An ammunition section responsible for the requisition, storage, issue, loading, and security of the ranger unit's basic load(s) of conventional munitions.

(6) A mess section provides all Class I support to the ranger unit. This includes operating a dining facility, providing food to remote sites, and drawing and issuing operational rations.

(7) A load team loads and unloads all equipment and supplies for the ranger unit. This team prepares pallets for loading, documents hazardous or dangerous cargo, and loads equipment pallets on USAF aircraft. It can also provide limited security for loads.

(8) A communications and electronics section provides secure communications among the RSE subelements, ranger unit, EOC, and DACG.

(9) An engineer platoon that is responsible for the construction of targets and rehearsal areas. This element of the RSE may or may not be activated, depending on the mission given the ranger unit. If the ranger unit deploys from CONUS to the objective area, this platoon has a key role to play in the premission preparation.

(10) A maintenance team whose primary mission is to ensure that the vehicles used by the RSE remain operational. It also provides predeployment maintenance and Class IX support for ranger unit vehicles and equipment.

(11) A medical team that provides unit-level medical care while the organic ranger medical section completes its outloading preparations. This team also provides any Class VIII supplies that the ranger unit may need.

(12) A security team responsible for security at the EOC, ranger unit location, and departure airfield. This team enforces stringent physical security precautions and maintains a constant counterintelligence (CI) surveillance of the outloading procedure. The security team provides armed guards for the protection of ammunition convoys, equipment movement, and vehicles.

(13) A rigger detachment that maintains, packs, rotates, and issues the personnel and equipment parachutes required by the ranger unit. It maintains enough stock of personnel parachutes to conduct a battalion-level parachute assault. The riggers lend technical skill to the rigging and aerial delivery of bulky, unusual, or delicate equipment.

b. The primary mission of the RSE is to provide the support necessary to quickly load and deploy the ranger units. The theater commander provides CSS to the ranger force once it has arrived at the OCONUS REMAB or ISB. The transition of support from the RSE to the CSS assets of the theater commander may be delayed for certain subelements in order to ensure a smooth deployment and enhance mission accomplishment. These subelements are released when the in-theater CSS assets can provide the support. Retained subelements of the RSE may be released from support requirements all at once, released in increments, directed to continue to provide support until mission completion, or augmented by in-theater CSS assets. The maintenance, medical, security, and rigger elements are those most likely to continue providing support OCONUS. This is likely if deployment is to an austere ISB pending immediate employment.

c. Once a ranger unit has been deployed OCONUS and has closed to a REMAB or ISB, the theater commander's staff coordinates CSS. If 1st SOCOM elements have deployed to the theater or operational area, the special operations support battalion (SOSB) of the 1st SOCOM coordinates CSS. Ranger logisticians at regiment and battalion level are mainly planners with limited logistics operations ability. Ranger logistical and personnel officers coordinate with CSS elements who provide such services.

d. The Theater Army Materiel Management Center (TAMMC) is the point of contact for information concerning in-theater logistical assets. In a developed theater of operations, a series of theater Army support groups will have been set up. One of these groups is assigned the mission of supporting the ranger force. The 1st SOCOM may also deploy certain assets available to provide CSS to the ranger force. The many CSS sources, the rapidly changing nature of SOF operations, and the unique CSS requirements of ranger units combine to make providing logistical and personnel support a challenge. The ranger unit's logistics officer must be included in the early phases of planning and deployment.

8-2. Logistics and personnel.

a. The echelons-above-corps CSS structure provides support to the ranger regiment as directed by the theater Army (TA) commander. In a low-intensity contingency involving a corps or less, there may be no echelons-above-corps logistics and personnel support elements. The corps support command (COSCOM) involved may be able to meet all needs from its own sources. However, in a combined operation in a high-intensity environment, the theater commander assembles a large CSS organization to provide support to the TA. This CSS base normally provides for all the ranger unit's requirements. During contingency operations in an area where there is no TA, the JTF commander provides the CSS elements. Until the buildup of these elements is complete, the ranger regiment must be prepared to operate from austere facilities with only the supplies it brings from CONUS.

b. The ranger regiment is normally assigned to the theater command for CSS. The TA commander, as directed by the theater commander, provides CSS to the ranger regiment and sets up support priorities. The EAC CSS organizations are under the command of the TA commander and consist of subordinate area and functional commands. The TA plans, coordinates, and executes its CSS functions through area-oriented support commands and specialized subordinate organizations. During the early stages of conflict, the TA may be organized as shown in Figure 8-1.

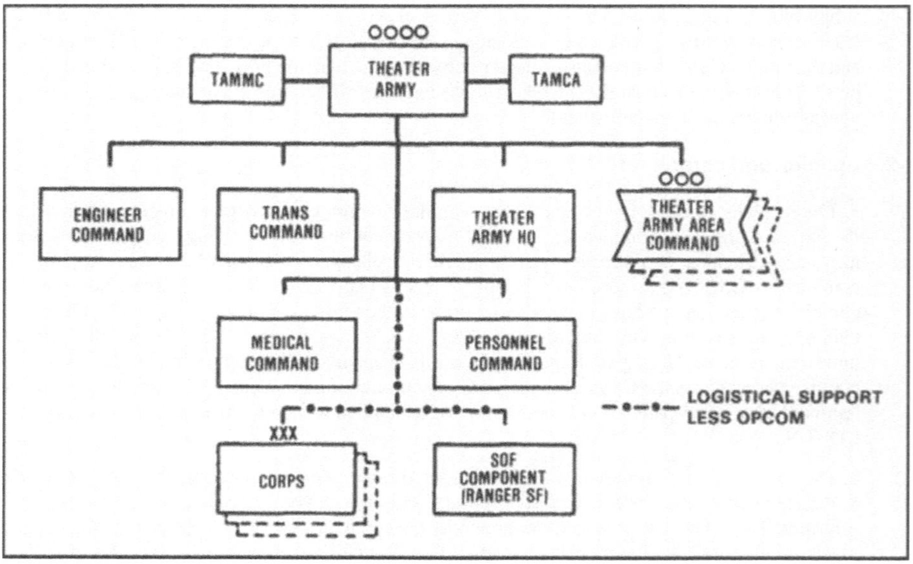

Figure 8-1. Ranger force relationship to theater Army.

c. The Theater Army Area Command (TAACOM) is the key CSS operator in the communications zone. It functions as a major subordinate command under the theater Army. The primary mission of the TAACOM is to provide direct support CSS to units such as the ranger regiment located in or passing through its assigned area. While located in an OCONUS REMAB, the ranger regiment receives most of its logistical support from the TAACOM.

d. The TAACOM area support group (ASG), which is subordinate to the TAACOM, is structured with operating units to provide direct support CSS to units located in or passing through its assigned geographical area of responsibility (see Figure 8-2). Area support groups are usually located along the lines of communication. The ASG is the prime source of CSS for the ranger regiment. It contains the units that actually provide services requested by the ranger unit logistics officer.

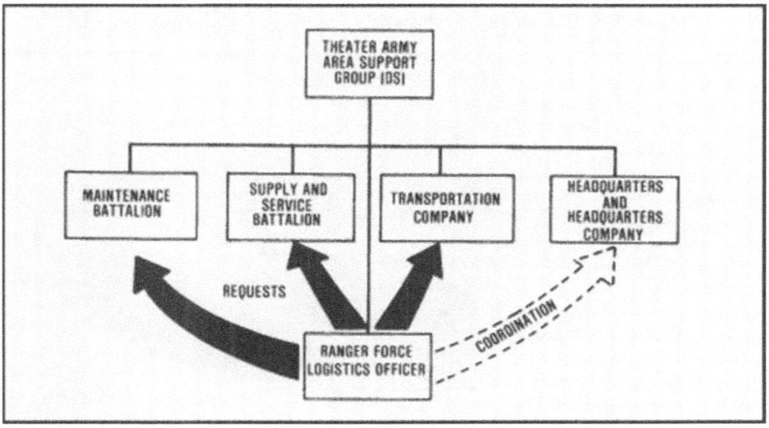

Figure 8-2. Ranger force relationship to theater Army area support group.

e. The ranger regiment is not normally assigned to a corps for employment. However, it may be placed under a corps' operational command for a specific mission. The ranger unit logistics and personnel officers must coordinate with the COSCOM that would then be providing CSS. (See Figure 8-3.) Corps support groups are major subordinate elements of the COSCOM. They provide command and control of COSCOM companies and battalions, which provide direct support (DS) and general support (GS) supply and field services to nondivisional units, such as the ranger battalion. (See Figure 8-4.)

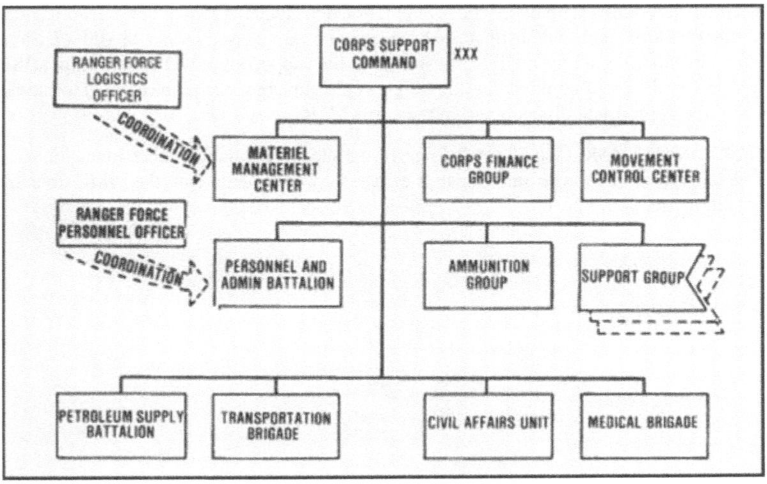

Figure 8-3. Ranger force relationship to corps support command.

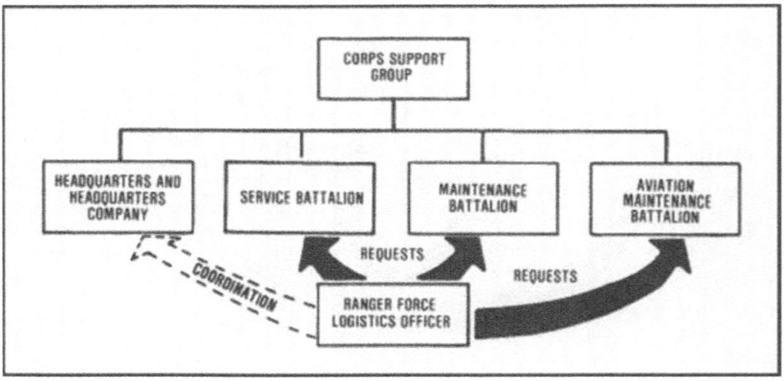

Figure 8-4. Ranger force relationship to corps support group.

f. Normally, the ranger regiment receives its logistical support from the corps support group responsible for the geographical area in which the REMAB or ISB is located. The corps support group provides all supplies except Classes V and VIII. The medical brigade provides Class VIII supply and medical maintenance support. The ammunition group provides Class V support. Financial and personnel services are provided by the corps finance group and the personnel and administration battalion, respectively.

g. Transportation needs are met by the transportation brigade. The ranger unit places ground transportation requests with either the Movement Control Center (MCC) or a designated movement control team. All airlift supply requirements are passed IAW FM 100-27. Standard and nonstandard supplies are normally requested from the supporting direct support unit (DSU). If authorized by the corps commander, the ranger regiment can place requirements for nonstandard items on the appropriate division within the corps MMC.

h. The COSCOM MMC (see Figure 8-5) performs supply and maintenance management for all classes of supply. The ranger unit logistics officer may coordinate with the MMC division chiefs or operations section.

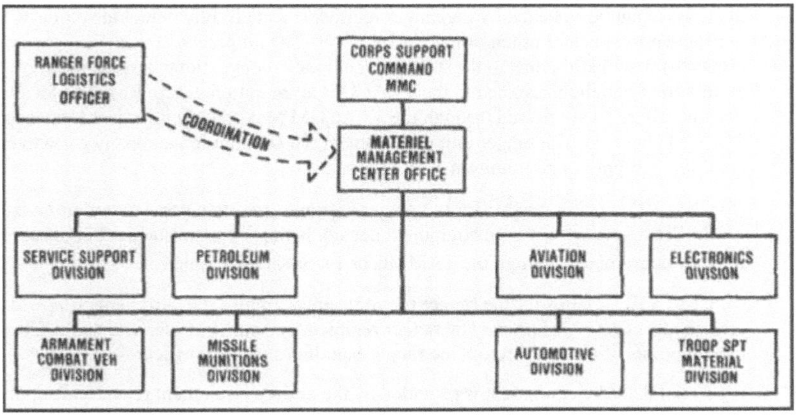

Figure 8-5. Ranger force relationship to COSCOM MMC.

i. Critical logistics operations, such as those supporting ranger and other SOF elements, need intensive management. They may be assigned to special staff sections that act as special operations force logistics managers. If designated, a logistics manager for special operations force elements would be a major point of contact for logistical support for the ranger regiment.

j. The ranger regiment may also receive logistical support from the SOSB. Normally, the SOSB provides administrative and logistical support to the Headquarters, Army Special Operations Forces (ARSOF). It complements the support provided by the Army component in the theater of operations by coordinating, augmenting, or assisting CSS forces. The SOSB provides unique quick-reaction support to the ARSOF or deploying SOF elements.

k. The ranger regiment does not make inordinate demands on the logistical system. However, a ranger unit may need certain nonstandard or low-density equipment. The concurrent need for nonstandard Class IX items requires management by logisticians to ensure support.

(1) Ranger battalion mess sections are responsible for Class I operations at the REMAB. The ranger regimental headquarters, which has no mess section, either collocates with one of the battalions or is supported by a mess facility operated by a nearby unit. It may share this facility on an area basis. The ranger unit personnel officer coordinates with the supporting DSU, which initiates the issue of Class I supplies.

(2) Class II, IV, and IX support to the ranger force is normally not of a great volume. It may consist of some nonstandard items. Depending on the need, the item, and the security, the SOSB may get some nonstandard items directly.

(3) Class III support needed by the ranger regiment is small and limited to standard items. The normal source of supply for Class III is the supply and service company providing area support.

(4) Class V support needed by the ranger regiment is small. It may include requests for nonstandard items of ammunition and explosives. The ranger logistics officer passes requests for standard items to the ordnance company (conventional ammunition) of the ammunition battalion assigned to the TAACOM ammunition group. Requests for nonstocked items are passed through the TAACOM Materiel Management Center or 1st SOCOM channels. The ranger unit must contact the ammunition company early to ensure a continuous supply of ammunition.

(5) Class VII supplies needed for the ranger regiment are often nonstandard or in excess of TOE. (See Appendix C for an equipment density listing.) All nonstandard or excess requests must be coordinated through the TAMMC or 1st SOCOM channels.

(6) Class VIII supplies for the ranger regiment are normally standard items supplied by the area support Class VIII point. The ranger regiment personnel officer sets up the Class VIII supply points of contact through the ranger battalion's medical officer.

(7) Aerial delivery equipment is provided by the airdrop equipment repair and supply company of the theater Army support group (TASG) (GS). There is normally one such company for each TAACOM. This unit supplies and repairs aerial delivery equipment. It does not pack and rig supplies for aerial delivery. The ranger regiment's requirements for aerial delivery supplies normally consist of standard items. Certain low-density items, such as HALO or HAHO parachute equipment, may also be provided through 1st SOCOM channels.

(8) The supporting motor transportation group of the transportation command is responsible for unit movement of the ranger regiment. The ranger regiment does not have any organic transportation assets. It depends on outside support whenever motor movement is needed.

8-3. Resupply.

a. Resupply operations for ranger units are normally covered during the planning and coordination phase of the mission. Ranger units try to complete their mission quickly and then be extracted or exfiltrate from the objective area. Ranger units carry into the objective area the equipment and munitions they need. They do not normally remain long enough to need extensive resupply. Certain operations may require the use of bulky or heavy items of equipment that cannot be carried by personnel. If the ranger force is airlanded on or near the objective, one or more of the aircraft may contain palletized supplies. These can be quickly off-loaded for future use.

b. If the ranger force is inserted into the objective area by parachute assault, aircraft can also drop resupply loads either just before or just after the personnel drop. These loads might include extra water, ammunition, demolitions, or barrier material. During peacetime, preloaded airdrop pallets are stored at CONUS depots for such contingency use. These mixed loads can be quickly moved to a departure airfield and inserted along with, or following, the ranger force. By dropping a resupply of water, food, equipment, and munitions with the ranger force insertion, the heavy-drop aircraft do not have to conduct a separate penetration of the enemy air defense zone. The ECM and SEAD programs used in the insertion of the personnel aircraft mask the penetration of the heavy-drop aircraft.

c. The aircraft dropping ranger personnel can also carry door bundles to deliver bulky or heavy items to the objective area. These door bundles, normally made up of A-7 and A-21 containers, may be color-coded to identify different loads or units. The most common door bundle loads are such items as 90-mm recoilless rifles and ammunition, 60-mm mortar ammunition, Dragon missiles, SHORAD missiles, or more demolition material. These bundles are pushed from the personnel aircraft just before the lead jumper's exit. They may be marked with lights or infrared sources for locating them at night.

d. If the tactical situation requires that a ranger force receive a resupply during the course of an operation, the most common method would be by airdrop. Airdrop could be needed if the operation takes longer than planned, initial supplies are lost, or changes occur in the operational plan. The ranger unit logistics officer and operations officers normally plan for resupply needs and schedule follow up loads of Classes I, II, IV, V, and VIII. There are two different airdrop requests channels for the ranger force to use:

(1) Nonemergency airdrop requests are sent through logistics channels. The ranger LOC sends requests for the airdrop of supplies and equipment to its supporting supply unit, and the supporting unit forwards the requests to the COSCOM or TAACOM MMC. The MMC coordinates with the movement control center (MCC). Jointly they direct the actions of the supply and transportation units. The MCC sends the request to the joint transportation board (JTB). The board validates it and assigns it a priority. The JTB then sends the request to the airlift control center for action. If the items requested are not on hand in the airdrop company, the MMC directs the proper supply activity to prepare the items for shipment to the airdrop unit. The MMC coordinates with the MCC, and the MCC ships the items. The airdrop unit prepares the supplies for the mode of delivery to be used. It also makes sure the supplies are delivered to the air terminal at the right time. (See Figure 8-6.)

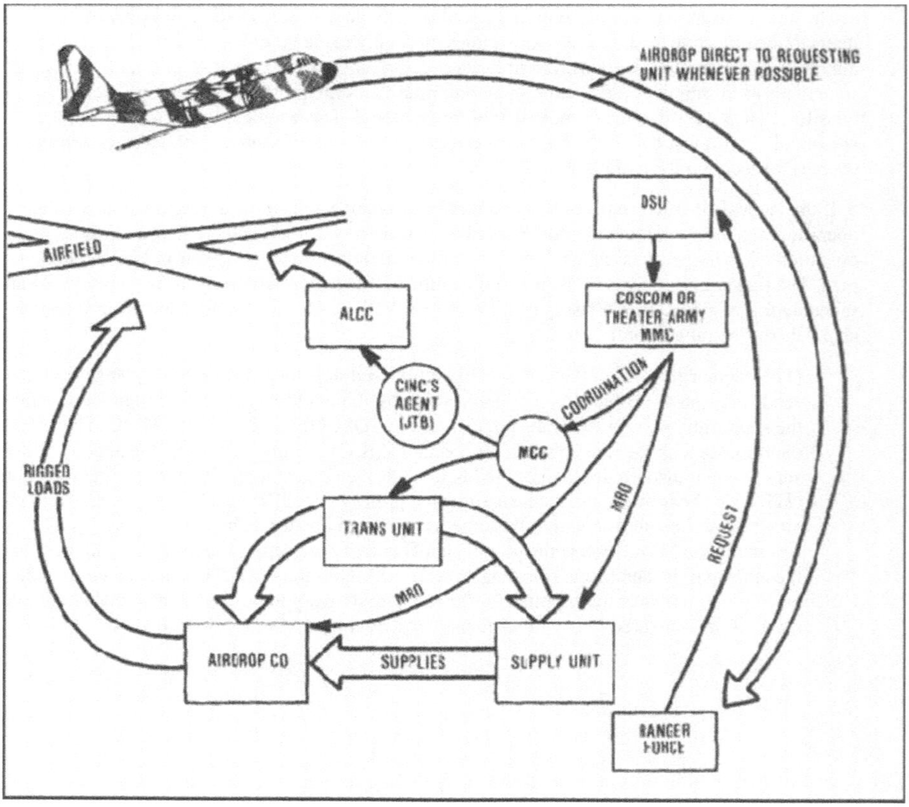

Figure 8-6. Ranger requests for nonemergency airdrop.

(2) Emergency requests for airdrop are passed through command channels in the quickest, most secure way.

(a) The request is usually sent to the command level that has authority to approve it, usually the tactical operations center at corps or above. The Air Force airlift officer at the direct air support center (DASC) coordinates with the Assistant Chief of Staff, Operations (G3) at the tactical operations center to determine whether airlift resources can be used for the mission.

(b) The TOC sends a request for an immediate airlift to the joint force commander's (JFC) designated agent for validation. At the same time, the airlift liaison officer advises the airlift control center (ALCC) that a request is coming through channels so plans can be made. When the ALCC receives the validated request from the JFC agent, it diverts or cancels less essential missions and provides airlift support. The TOC sends the request for supplies or equipment to the MMC.

(c) The MMC and MCC jointly direct the actions of the supply and transportation elements. The airdrop company stocks Classes I, III, and V supplies. It holds these supplies for emergency missions. If the supplies and equipment requested are not on hand in the airdrop company, the MMC directs the proper supply activity to prepare the items for shipment to the airdrop unit. The MMC coordinates with the MCC to ship the items. The airlift is arranged by the ALCC.

(d) The airdrop unit prepares the supplies according to the mode of delivery to be used. It coordinates with the MCC and delivers the supplies to the designated point at the air terminal (usually the cargo aircraft ramp). (See Figure 8-7.)

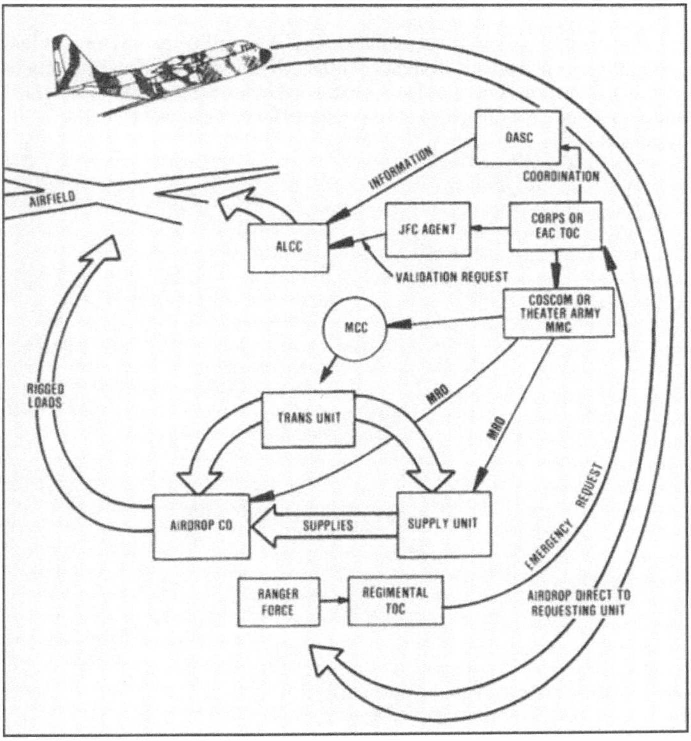

Figure 8-7. Request channels for emergency airdrop.

e. Aircraft delivering airdrop resupply deep behind enemy lines need to take precautions to avoid enemy detection and antiaircraft fire. The USAF is responsible for most of these precautions, such as ECM, suppression of enemy air defense (SEAD), and WILD WEASAL support. These do not concern the ranger commander. The choice of the type of parachute delivery system to use, however, may affect the ground commander's plan. The ranger force commander must specify any tactical restrictions that would require a specific type of delivery system. The ranger LOC and regimental logistics officer must advise the ranger force commander on the type of parachute system to be used. Conflicts must be quickly resolved so as not to interfere with the tactical operations.

(1) The safest way for the airdrop aircraft to penetrate enemy air defense and remain undetected is often by use of very low-level flight. There are four low-level parachute delivery systems that can be used without the aircraft being required to climb to a higher drop altitude and risk detection by enemy ADA.

(a) The low-altitude parachute extraction system (see Figure 8-8) needs a length of level ground with enough clearance for the delivery aircraft to make a very low pass over the extraction zone. The load is extracted by parachute from the aircraft and slides to a stop on a rugged, skid-type pallet. Heavy loads can be delivered using this system.

Figure 8-8. Low-altitude parachute extraction system.

(a) The containerized delivery system (CDS) (see Figure 8-9) consists of multiple, individually rigged A-22 containers, each with its own parachute, weighing up to 2,000 pounds each. The system provides single-pass delivery of up to 16 containers by C-130 aircraft and up to 28 containers by C-141 aircraft. The CDS loads may be delivered into drop zones using multiple points of impact to allow for tactical separation. The ranger force commander can direct the use of multiple points of impact, noting the advantages to be gained against the requirement for multiple passes.

Figure 8-9. Containerized delivery system.

(c) The high-speed, low-level, airdrop system (HSLLADS) (see Figure 8-10) is a single A-21 container specially rigged to withstand the parachute opening shock when airdropped from C-130 aircraft at high speed. This system can be used to deliver up to 600 pounds for each container, with a maximum of four containers for each pass, at speeds up to 250 knots.

Figure 8-10. High-speed, low-level, airdrop system.

(d) The CTU-2/A is a high-speed, aerial-delivery container (see Figure 8-11) that can be used to deliver supplies from high-performance aircraft flying at a minimum altitude of 300 feet above ground level and a maximum airspeed of 425 knots. The CTU-2/A is carried on the bomb racks of fighter or bomber-type aircraft. Upon release, a pilot parachute deploys the main chute and the container descends slowly.

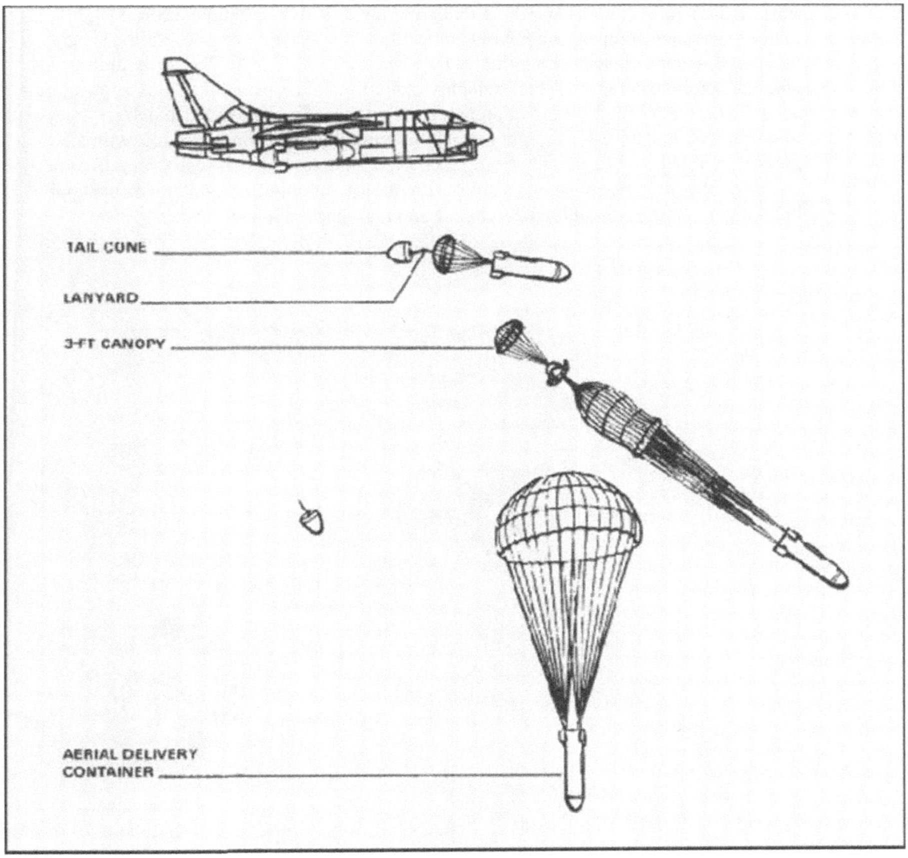

Figure 8-11. CTU-2/A high-speed, aerial-delivery container.

The CTU-2/A can be used to deliver up to 500 pounds of supplies such as weapons, water, food, or munitions. The container can be destroyed by burning. The main advantage of this system is that it can be delivered by high-performance aircraft deep behind enemy lines and in a dense air defense environment. The accuracy of this system is equal to that of a conventional bomb strike.

(2) Sometimes it may be better to drop resupply loads to the ranger force from a high altitude. The HAARS (see Figure 8-12) permits containerized unit loads weighing from 200 to 2,000 pounds to be delivered from aircraft at speeds up to 150 knots from up to 25,000 feet AGL. The HAARS consists of a cargo parachute, an airdrop container, an altitude

sensor, and a pilot chute. The pilot chute gives the descending bundle a speed slightly greater than an accompanying HALO parachutist. The system provides for the steady free-fall descent of loads from altitudes between 2,000 and 25,000 feet to an altitude at which a barometric sensor actuates deployment of the main parachute. This allows safe and accurate delivery of loads onto unprepared drop zones. The HAARS can deliver a payload to within 260 meters from a target impact point from a 10,000-foot altitude, with a proportional degree of accuracy from 25,000 feet. Ranger units can use the system for resupply of battalions and smaller units with rations, ammunitions, and medical supplies, breaking down the containerized material into man-packed loads.

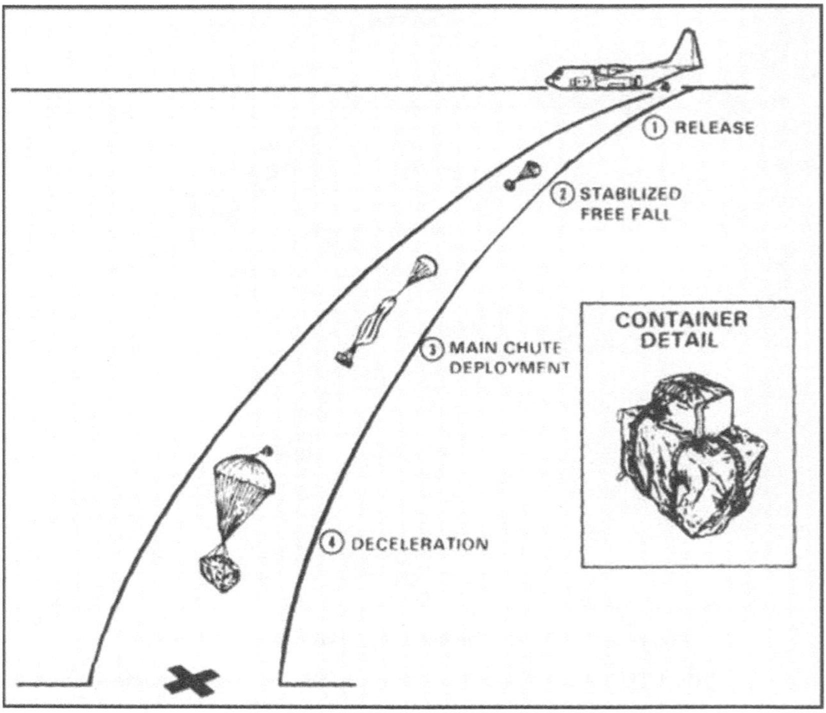

Figure 8-12. High-altitude, airdrop resupply system.

(3) The ranger force may need aircraft loads delivered during periods of poor visibility. The adverse weather aerial delivery system allows aircraft to drop resupply bundles accurately in adverse weather or darkness. These AWADS operations can be conducted safely and effectively with a minimum 300-foot ceiling AGL and a minimum visibility of less than half a mile. The minimum-drop altitude is 500 feet above the highest obstruction. These minimum ceiling or visibility restrictions do not apply to the Combat Talon aircraft.

f. Resupply to the ranger force can be provided by Army helicopters (see Figure 8-13). Normally this is done by SOA units. The danger to helicopters during long-range, low-level penetration of enemy rear areas requires that resupply be conducted when there is limited visibility.

Figure 8-13. Resupply bt helicopter.

8-4. Military police.

a. The ranger regiment needs military police support to provide security at the REMAB. Although ranger units can provide their own security and local defense, using military police for this purpose allows the ranger unit to concentrate its manpower and training on mission preparation. A military police security company is normally designated to provide support to 1st SOCOM elements located in the communications zone. Military police provide security patrols of the REMAB, convoy security, and vehicular search. They provide entry and exit control of the sensitive compartmented information facility (SCIF).

b. A preemptive strike by enemy special operations forces is a threat to the ranger regiment while located in the REMAB; either by conventional attack or by terrorist-type bombing. The preparation and location of the headquarters, the type of buildings, the location of landing zones, the traffic flow, and the size of the local population all influence the size of the military police unit. A military police platoon, augmented with bomb and guard dog teams, should be enough security at the REMAB. These security forces would not normally deploy from the REMAB to an ISB with the ranger unit.

8-5. Counterintelligence.

Counterintelligence support to the ranger regiment is normally provided by a CI team from the military intelligence group at TA level. The CI team provides a liaison and informational flow from the CI agencies operating within the theater. It also acts as advisor to the ranger unit commander on OPSEC while performing CI surveillance operations in the vicinity of the REMAB. The CI support may also be provided through 1st SOCOM channels. It is through 1st SOCOM that liaison with national intelligence and counterintelligence agencies is achieved.

8-6. Personnel service support.

The personnel service support needs of a ranger unit are much like that of any other unit of the same size. To receive the support his unit needs, the ranger personnel officer must meet with the proper agencies within the theater of operations.

a. **Strength accounting.** Within the ranger regiment, strength accounting is conducted through normal personnel channels. Regimental strength figures are passed quickly to the TAACOM or COSCOM. These figures are crucial to the theater commander for planning future operations. When matched with casualty reports, they provide the information needed for replacements.

b. **Replacement operations.** Ranger units need exceptionally well-trained and physically conditioned personnel as replacements. Ranger operations require that replacements train with the unit before they can become an effective member of the team. There are five sources of replacements for ranger units:

(1) **In-theater volunteers.** Use of in-theater volunteers, preferably with experience in ranger units, is the quickest way to replace personnel. These volunteers have a knowledge of the basic skills required of rangers and have some operational experience. The ranger regimental personnel officer coordinates with the theater personnel officer to screen and select those volunteers with the required MOSs and grades. Volunteers from other special operations force components, or from other theaters of war, are identified by DA and screened for assignment to the ranger regiment. Even though these volunteers have individual training and experience, it takes several weeks to become fully integrated into the ranger unit's operations and to achieve the needed skill level.

(2) **Reorganization of units.** This method of reconstitution is used when two or more ranger units have high personnel losses. The remaining members of these units are reorganized into a single unit. This method provides the ranger regimental commander with enough ready-trained, physically fit personnel to form a full-strength unit. Extensive unit training must be conducted before such a reorganization regains the unity and esprit of its component parts. The high standards found among the subelements of the ranger regiment makes this a good method of reconstituting a ranger unit. The units that have lost all their personnel may be reconstituted later, using some other method, or left unreconstituted,

depending on the situation.

(3) **Hospital returnees.** These personnel, former members of the ranger regiment who were injured, evacuated, treated, and have now recovered, are valuable assets. They provide the unit with a major reconstitution ability. These personnel should be returned quickly to their original battalion. This enhances unity and increases morale and esprit. Returnees may not be physically able to return to full duty right away. While recuperating, they may be assigned to limited duty as instructors. The ranger unit uses the returnee's combat experience and individual skills to screen and train newly assigned volunteers. Lessons learned in combat can help prepare the unit for future operations.

(4) **Replacement volunteers from CONUS.** These personnel have training in the basic skills and may have completed the ranger course, but they do not have any ranger unit experience. They need several weeks of unit training to become prepared for combat operations.

(5) **Ranger indoctrination program.** During wartime, this program provides screening and training of newly assigned personnel. Volunteers who have neither served in a ranger unit before nor have graduated from the ranger course require extensive individual training before they are ready to begin unit training with a ranger battalion. This method of reconstitution of ranger units requires the longest training period. During this reconstitution phase, a ranger battalion needs extensive use of ranges, maneuver areas, and support by aviation and other special operations-type units. The ranger regimental commander may choose to move the program from CONUS to a location near the regiment's REMAB.

c. **Casualty reporting.** Casualty reports are handled through regimental personnel channels and must be timely and accurate. The reports are used to check strength accounting data, to assist graves registration personnel in the recovery and identification of remains, and to notify next of kin.

d. **Health service.** The health service support needed by the ranger units deployed OCONUS is usually small. Area medical support provided by host country or theater medical command assets usually meets their needs. Evacuation of wounded members of the ranger force begins upon extraction from the objective area. Personnel wounded in action are evacuated through USAF or Army channels to the nearest medical facility that can provide treatment. The ranger force medical officer and personnel officer coordinate with the area medical support facility providing preventive and routine medical care to the unit while at the REMAB. Vaccine needed for innoculation of the ranger force members and any special Class VIII supplies needed for the mission are drawn through normal medical channels.

e. **Financial service.** Financial support varies with the situation. In mid- and high-intensity conflicts, regular paydays are suspended in the theater of operations. As a nondivisional unit, the ranger regiment coordinates with the theater finance center or corps finance group for financial support. In the REMAB, ranger units are provided financial support from the closest finance support unit. The ranger unit personnel officer appoints Class A agents to help with the disbursement of funds. The ranger force may be given a specific amount of local currency or military scrip depending on the length of time in the REMAB and the operational situation. The ranger unit personnel officer also coordinates with the theater finance center to get any third-country currency that may be included as part of an escape-and-evasion kit provided to ranger elements.

f. **Postal service.** Postal service to the ranger unit is through normal channels. Mail for members of the deployed force is forwarded by the rear detachment commander to an established theater Army general support postal activity. This activity forwards the mail to the postal unit designated to support the REMAB site. The ranger unit commander appoints subordinate unit officers as censors. The exact location of the REMAB, and all aspects of the mission and its location, remains classified until released by an official Army spokesman. The ranger unit commander must ensure that the forwarding of mail, or the writing home of personal letters, does not endanger the ranger force mission. All outgoing mail is censored during wartime. For security reasons, outgoing mail may be held for a short time before it is forwarded. Mail from the ranger unit is never sent by host country or other nation's postal services.

g. **Chaplain activities.** Chaplain activities are provided by the chaplain assigned to the ranger battalion. He provides unit coverage through services, rites, and sacraments regardless of his or the unit member's religious affiliation.

h. **Legal service.** Ranger units do not need extensive legal service support. Legal service support is obtained through coordination with the theater Army staff judge advocate.

APPENDIX A

Organization and Equipment Introduction of the Ranger Regiment

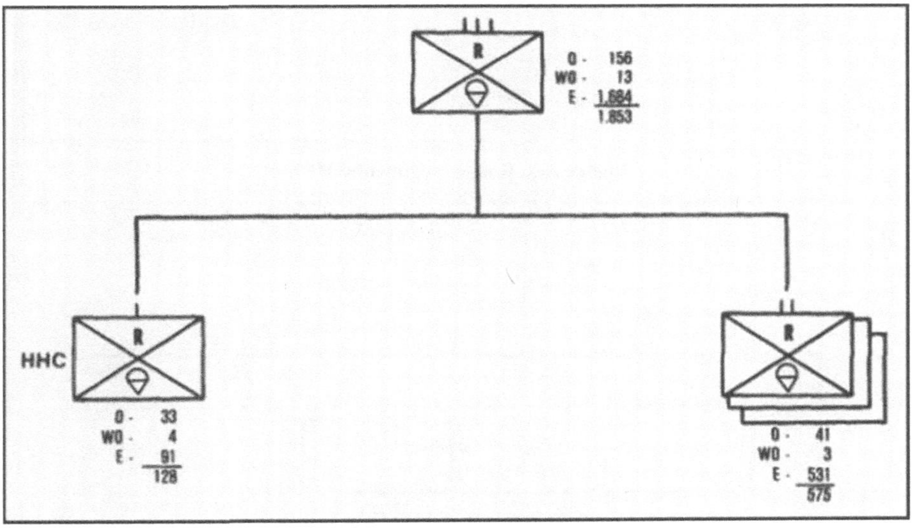

Figure A-1. Ranger regiment.

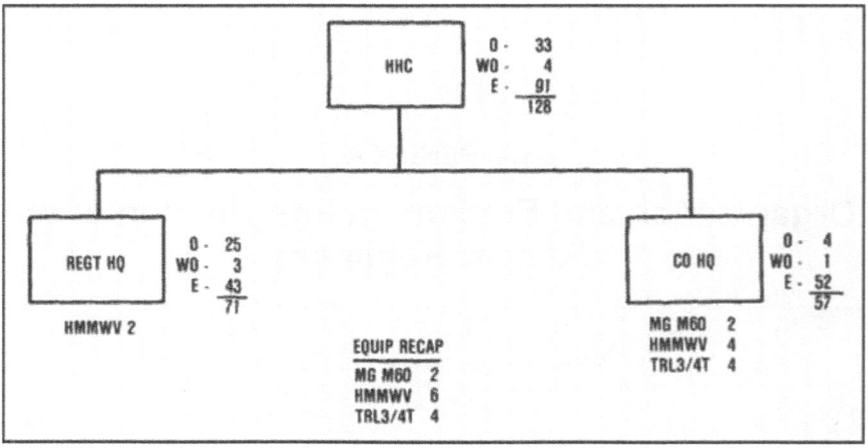

Figure A-2. Ranger regimental HHC.

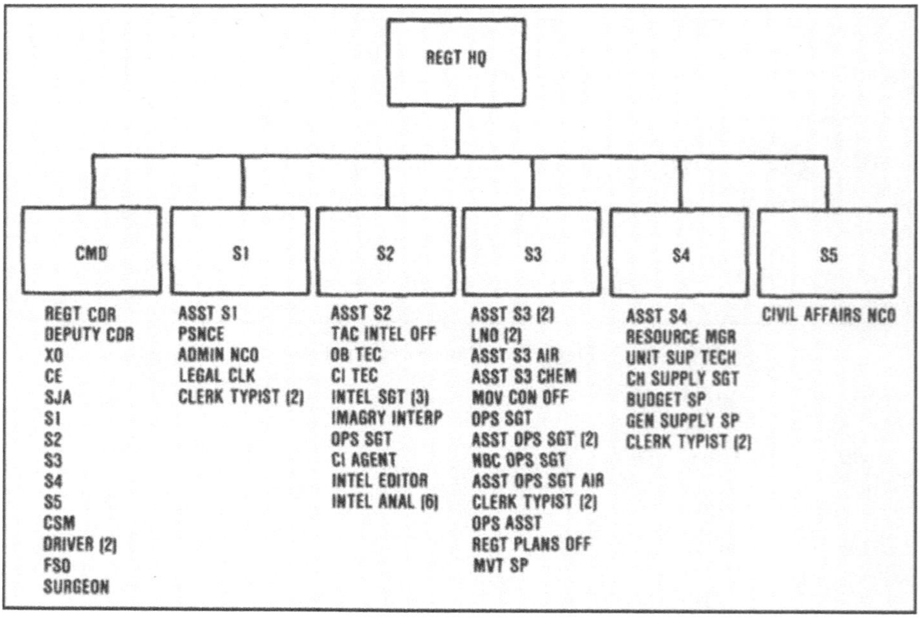

Figure A-3. Ranger regimental staff.

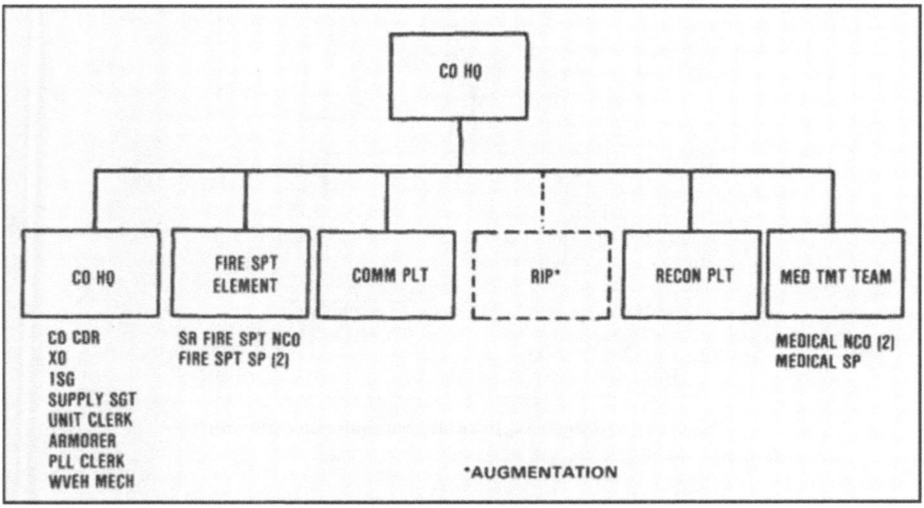

Figure A-4. Ranger regimental headquarters company.

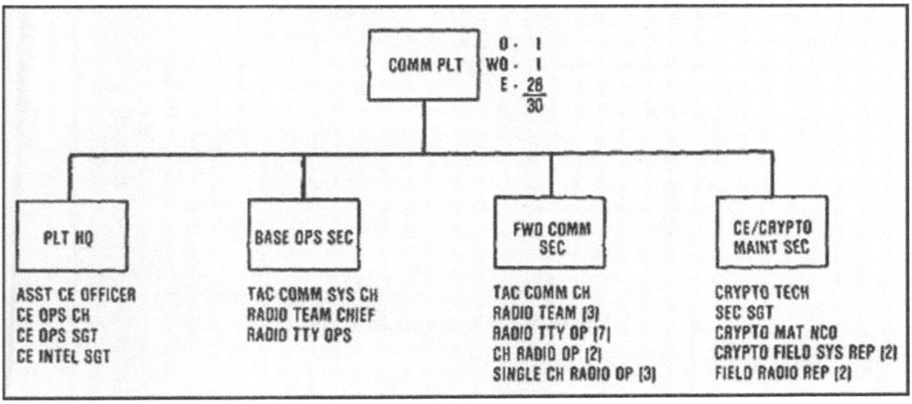

Figure A-5. Ranger regimental communications platoon.

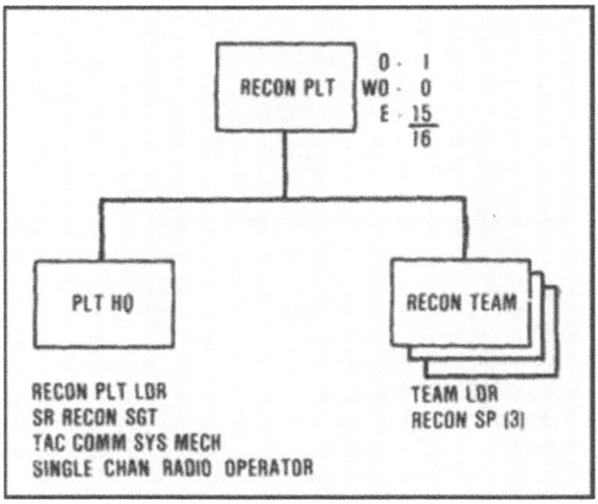

Figure A-6. Ranger regimental reconnaissance platoon.

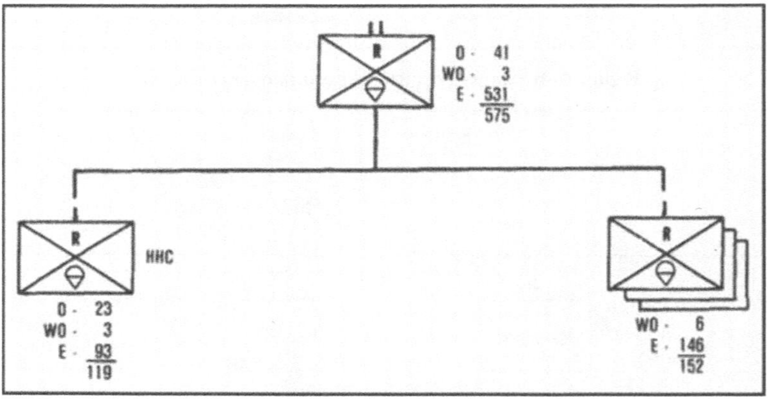

Figure A-7. The ranger battalion.

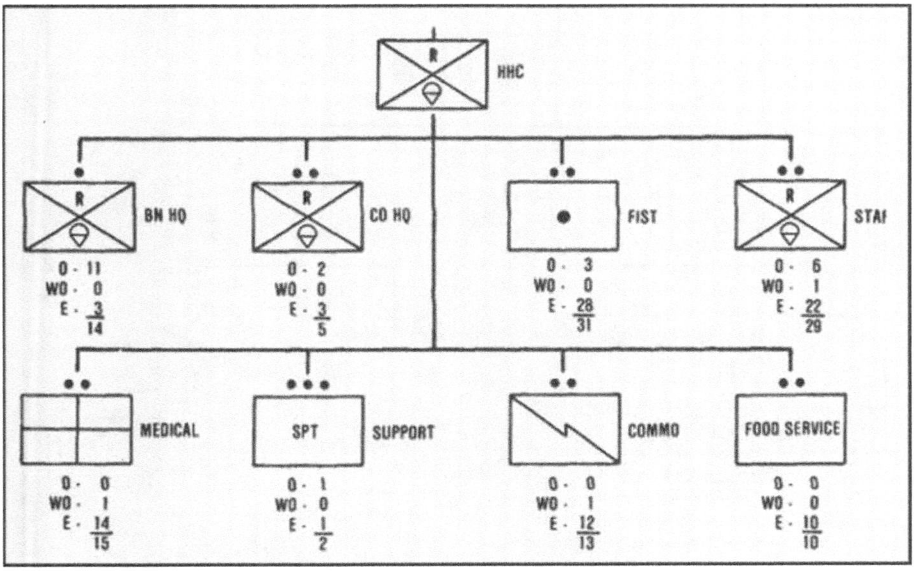

Figure A-8. Ranger battalion HHC.

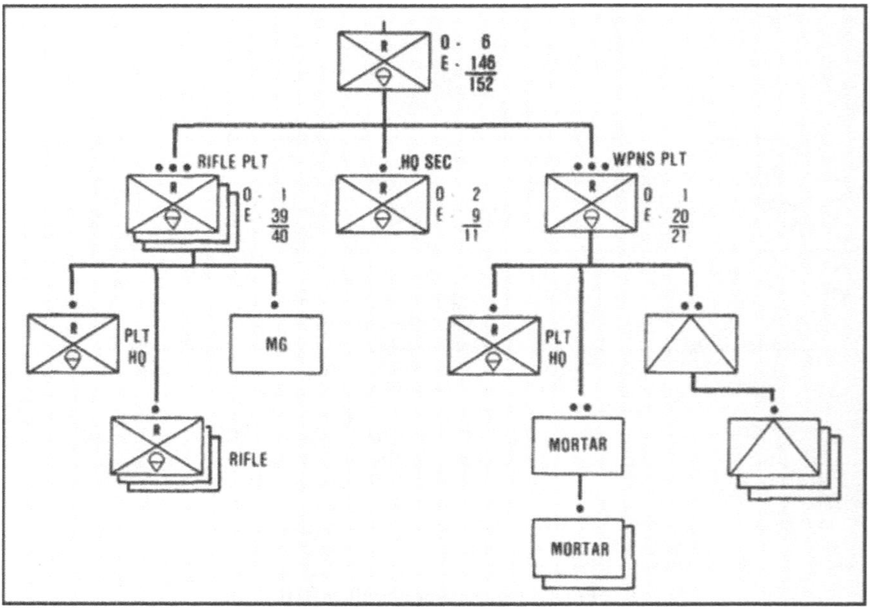

Figure A-9. Ranger rifle company.

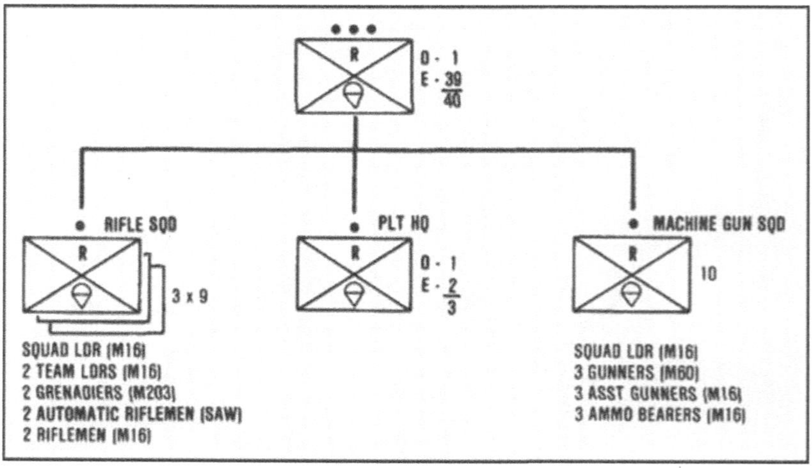

Figure A-10. Ranger rifle platoon.

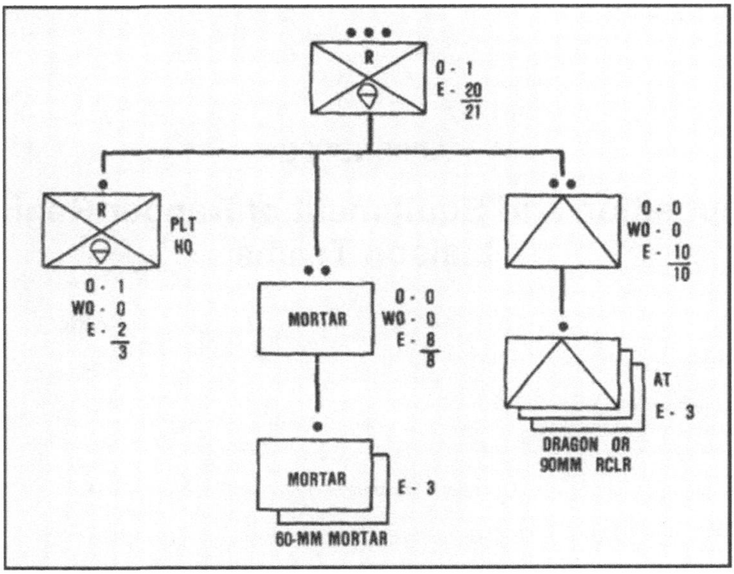

Figure A-11. Ranger weapons platoon.

APPENDIX B

Composition and Equipment of Ranger Regiment Liaison Teams

Team Red	**RANK**
Liaison Officer	MAJ
Senior Intelligence Analyst	SSG
Clerk-Typist	SP4
Tactical Communications System Operator/Mechanic	SGT
Combat Telecommunications Center Operator	SP4
Tactical Communications System Operator/Mechanic	SP4
Team White	
Liaison Officer	CPT
Senior Intelligence Analyst	SSG
Clerk-Typist	SP4
Tactical Communications System Operator/Mechanic	SGT
Combat Telecommunications Center Operator	SP4
Tactical Communications System Operator/Mechanic	SP4

Equipment (for Each Team)

HMMWV with 3/4-ton trailer	5-kw generator
Communications shelter	Tent, GP, small
AN/PCS-3 radio set (3)	AN/VSC-7 net control station (1)
MX 300 (4)	KY 57 with battery case (5)
HYX 57 with battery case (4)	HYP 57 (5)

KYK 13 (2)

KY 65 with battery case (1)

KG 84 (1)

H189 (2)

GRC 193 (1)

VGC-74 (1)

PP 1648 (1)

SB-22 switchboard (1)

KYI 18 (1)

KY 65 WLA (1)

Digital message device (1)

AN/PRC-77 radio (1)

Facsimile transmission device (1)

TA 312 with WD1 (5)

APPENDIX C
Equipment Density

C-1. Communication equipment.

	Regimental Headquarters	Battalions (x3)
AN/VSC-7 net control station	7	2
AN/PSC-3 radio set	23	13
AN/PRC-41 radio set	2	5
AN/PRC-104 radio set	9	26
AN/GRC-193A radio set	4	0
AN/PRC-70 radio set	6	0
AN/PRC-77 radio set	22	50
AN/VRC-46 radio set	2	0
AN/VRC-49 radio set	0	2
AN/PRC-68 radio set	0	66
TSEC/KY-57 speech security device	60	68
TSEC/KY-58 speech security device	6	0
TSEC/KY-65 speech security device	11	0
AO-8990/P digital message device group	15	23
AN/GXC-7A tactical FAX device	7	2
AN/UGC-74A(V)3 communications terminal	7	0
SB-22 switchboard	2	0
SB-3614(V)/TT switchboard	2	1

TSEC/KG-84	5	0
SB-993/GT switchboard	0	12

C-2. Weapons and equipment each battalion.

Infrared aiming light	96	Interrogator set, Stinger missile	3
Rifle, M16	451	Grenade launcher, M203	67
Rocket launcher, M202	13	Machine gun, M60	27
Mortar, 60-mm M224	6	Rifle, sniper, M21	24
Recoilless rifle, 90-mm	18	Viewer, infrared	9
Tracker, Dragon	9	Night vision devices (all types)	322
Truck, 1/4-ton	2	Laser target designators	11
Radio (all types)	164	Tactical satellite radios (all types)	15
Squad automatic weapon	54		
Laser range finder	14		

APPENDIX D

Standing Orders, Rogers' Rangers

1. Don't forget nothing.
2. Have your musket clean as a whistle, hatchet scoured, sixty rounds powder and ball, and be ready to march at a minute's warning.
3. When you're on the march, act the way you would if you was sneaking up on a deer. See the enemy first.
4. Tell the truth about what you see and what you do. There is an army depending on us for correct information. You can lie all you please when you tell other folks about the Rangers, but don't never lie to a Ranger or officer.
5. Don't never take a chance you don't have to.
6. When we're on the march we march single file, far enough apart so one shot can't go through two men.
7. If we strike swamps, or soft ground, we spread out abreast, so it's hard to track us.
8. When we march, we keep moving till dark, so as to give the enemy the least possible chance at us.
9. When we camp, half the party stays awake while the other half sleeps.
10. If we take prisoners, we keep 'em separate till we have had time to examine them, so they can't cook up a story between 'em.
11. Don't ever march home the same way. Take a different route so you won't be ambushed.
12. No matter whether we travel in big parties or little ones, each party has to keep a scout 20 yards ahead, 20 yards on each flank, and 20 yards in the rear so the main body can't be surprised and wiped out.
13. Every night you'll be told where to meet if surrounded by a superior force.
14. Don't sit down to eat without posting sentries.
15. Don't sleep beyond dawn. Dawn's when the French and Indians attack.
16. Don't cross a river by a regular ford.

17. If somebody's trailing you, make a circle, come back onto your own tracks, and ambush the folks that aim to ambush you.

18. Don't stand up when the enemy's coming against you. Kneel down, lie down, hide behind a tree.

19. Let the enemy come till he's almost close enough to touch. Then let him have it and jump out and finish him up with your hatchet.

MAJ Robert Rogers

APPENDIX E

The Ranger Creed

R Recognizing that I volunteered as a Ranger, fully knowing the hazards of my chosen profession, I will always endeavor to uphold the prestige, honor, and high "esprit de corps" of the Ranger Regiment.

A Acknowledging the fact that a Ranger is a more elite soldier who arrives at the cutting edge of a battle by land, sea, or air, I accept the fact that as a Ranger, my country expects me to move farther, faster, and fight harder than any other soldier.

N Never shall I fail my comrades. I will always keep myself mentally alert, physically strong, and morally straight and I will shoulder more than my share of the task, whatever it may be. One hundred percent and then some.

G Gallantly will I show the world that I am a specially selected and well-trained soldier. My courtesy to superiors, my neatness of dress, and my care of equipment shall set the example for others to follow.

E Energetically will I meet the enemies of my country. I shall defeat them on the field of battle, for I am better trained and will fight with all my might. Surrender is not a Ranger word. I will never leave a fallen comrade to fall into the hands of the enemy and under no circumstances will I ever embarrass my country.

R Readily will I display the intestinal fortitude required to fight on to the Ranger objective and complete the mission, though I be the lone survivor.

RANGERS LEAD THE WAY

APPENDIX G

Training

G-1. Application and guidance.

a. To prepare for combat, ranger units train on missions derived from their analysis of operational and contingency plans and the ranger mission training plans (MTPs). Training guidance is in doctrinal manuals, field circulars, and soldier's training publications. Training must ensure that the ranger unit is skilled in all basic light infantry skills and in ranger-unique skills. In the final analysis, good training depends upon the leadership and teamwork of ranger unit officers and NCOs.

b. Everything rangers do should prepare them to perform well in combat. Ranger units train IAW the battalion training management system (BTMS) concentrating on basic skills and on live-fire, hands-on training.

c. An effective training program breeds high morale and few discipline-related problems. Some qualities of ranger units that have good training programs are as follows:

(1) Leaders who can solve problems.

(2) A thorough and continuous analysis of the unit's mission

(3) Validated training objectives.

(4) Attention to the basics (move, shoot, and communicate).

(5) Multiechelon sustainment training.

(6) Standardized procedures.

(7) Physical and mental strength.

(8) Live fire as an integral part of training.

(9) Safety as a basis to all operations.

(10) The use of night and adverse weather and difficult terrain as combat force multipliers,

d. When all training is planned with realism, rangers will be in training situations that require the same attention as in battle. Realism and stress in training result in faster learning and better retention of skills. Training objectives must be clear with measurable standards to determine proficiency. Rangers and ranger units must train to the standards that will be expected in combat.

e. Examples of unit training goals include:

(1) Developing each member of the chain of command to perform his duties with competence, confidence, and pride.

(2) Maintaining small unit integrity during all training and support activities. Accomplishing tasks as units.

(3) Ensuring that all tasks for training are keyed to mission needs and weaknesses. Assets--ranger's time, money, and materiel--are expended only after thorough planning.

(4) Training to standard.

(5) Developing and maintaining discipline and esprit to ensure the abilities of each ranger and all units are achieved.

(6) Emphasizing use of the ranger mission training plan (MTP) for each echelon of the organization.

(7) Emphasizing the basics.

(8) Maintaining a high level of physical fitness.

(9) Working together within the ranger regiment and other elements of the special operations forces community.

(10) Developing a unit atmosphere of quick reaction under stress. Encouraging and rewarding individual initiative. Accomplishing unit missions IAW the commander's intent.

G-2. Philosophy.

a. To be effective, training should be physically and mentally challenging. Each ranger will then have a sense of purpose, accomplishment, and satisfaction. Training is evaluated by the results rather than hours of training. Units that are well trained today need reinforcement training to sustain readiness. Training management must be continuous. Training managers must use the time available to improve training efficiency. Cyclic training will not produce the desired results. The "train-to-correct deficiencies" method is needed. The training managers must analyze weaknesses, conduct prescriptive and reinforcement training, and reevaluate their unit's training status. The process that is used to develop and maintain a continuous training and evaluation program is shown in Figure G-1.

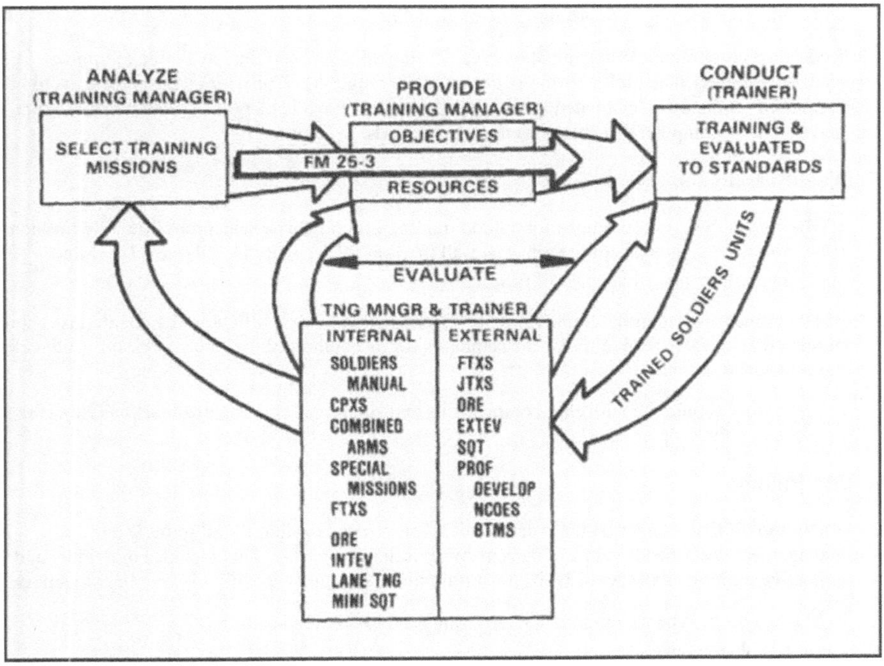

Figure G-1. The Army training and evaluation program.

b. Individual and collective training must be integrated within a unit training program. Training of the individual ranger, leaders, and units at each level must occur at the same time. The ranger commander or leader plans and conducts training for every member of his unit.

c. Ranger commanders analyze operational and contingency plans to determine unit missions and translate the associated tasks into training objectives. These objectives form the basis of the unit training program. The regimental headquarters provides training guidance in a formal regimental training plan. Commanders at lower levels program, supervise, and conduct training within their units.

d. All members of the chain of command are responsible for training. The NCO leader or first-line supervisor conducts individual and small unit collective task training. The officer plans individual training in his unit and directs and conducts collective training.

e. Centralized development of training objectives and decentralized training builds well-trained units and outstanding leaders. Authority and responsibility for the detailed planning, scheduling, conducting, evaluating, and supervising of training are delegated to the lowest element capable of managing this training. Higher headquarters is responsible for establishing objectives, monitoring and evaluating training, and issuing general guidelines.

f. Multiechelon training and evaluation means using all assets and time available to improve combat readiness. The multiechelon structure of the ranger MTP provides a basis for units to be continually trained and evaluated by internal evaluation. Each leader must know his mission tasks and train to accomplish those tasks to the set standards.

G-3. Mission analysis.

a. Mission analysis is continuous for a good training program. The selection of training missions or tasks results from analysis of the unit's overall mission. (See Figure G-1 above.) Tasks and subtasks are then placed in order of priority.

b. The trainer must identify countertasks for the opposing force (OPFOR). The countertasks should be such that an OPFOR that meets the standards for its countertask will be successful. Both sides have a chance to win.

c. The trainer should then develop combat drills and situational training exercises (STXs), if not furnished by the proponent school.

G-4. Unit training.

a. **Responsibility.** Ranger commanders at all levels are responsible for all training and management described herein and current Army training doctrine. This includes individual and collective training at all levels. Units must train to the standards specified in training objectives.

b. **Management.** The training management model for commanders is:

- Set objectives.
- Provide resources.
- Coach subordinates.
- Measure results.
- Provide feedback.

G-5. Individual training.

There are three elements basic to a well-trained individual ranger. He must be physically fit, be proficient in the use of his weapon, and have the psychological commitment to fight the enemy.

a. Individual training is the responsibility of the first-line supervisor under the supervision of the chain of command. It is conducted using performance-oriented techniques under battlefield conditions. New skills must be developed and practiced, and acquired skills must be reinforced throughout the training year. A high level of proficiency must be maintained through continuous training and practice.

b. Trainers should use peer instruction. Commanders ensure needed resources are available to conduct challenging individual training. Ranger units schedule and conduct this training and evaluate individual proficiency.

c. Flexibility in planning and scheduling is necessary. The goal of all individual training must provide for reinforcement training to offset the effects of the forgetting curve. (See Figure G-2).

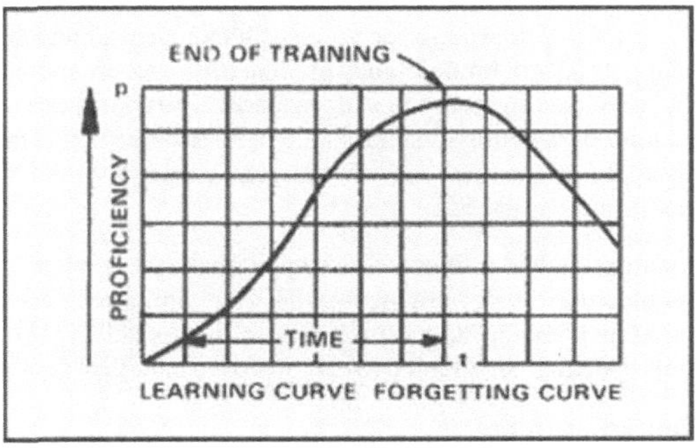

Figure G-2. The learning curve and forgetting curve.

d. The command sergeants major, unit first sergeants, and staff section sergeants are the keys to an effective training program. These senior NCOs must counsel, guide, and critique junior NCOs in training preparation and execution. Fire team leaders, section chiefs, and squad leaders coordinate and present individual training.

e. The NCO supervisors must ensure that the ranger is trained in the skills required for the specific military occupational specialty (MOS) and in those skills used daily. Not all training can be acquired in an office. For example, all rangers must maintain a high skill level with their weapons, be trained in NBC subjects, and be capable of performing contingency missions.

G-6. Collective training.

a. **Guidance.** Collective training trains crews, teams, squads, platoons, companies, and battalions to do their missions. The ranger MTP is used to plan training and evaluate the unit's weaknesses and strengths. If a ranger commander receives new training requirements or missions, they should become part of the collective training program. The ranger MTP describes the combat tasks to be performed by every echelon of the ranger battalion. It is a plan for organizing, conducting, and evaluating the year-round unit training program.

b. **Responsibility.** The ranger unit commander is responsible for the conduct of collective training and for all other training in his unit. Practical exercises emphasize the techniques of terrain use, movement, and effective employment of weapon systems. Training under stress will be conducted to move the unit to a higher plane of combat readiness.

c. **Drills.** Drills are written for small units--for example, fire team and squad, crew and section--and describe an action that is shorter than an entire STX. Drill training objectives cover both individual and collective skills (critical collective tasks) and list them together for effective group action. Drills must be mastered in order to react to rapidly changing situations in combat. The trainer can interrupt the execution of a drill to provide necessary training on individual or collective skills.

d. **Situational training exercise.**

(1) The STXs are short, mission-oriented tactical exercises in which a group of closely related collective tasks are trained. They are like drills but are more complex and flexible. They usually involve a larger unit than does a drill. They integrate leader training, leader tasks, drills, and separate individual tasks into a realistic setting. Trainers may execute STXs as written or tailor them to meet specific unit requirements. An STX is not standardized, but it must have a firm doctrinal base.

(2) Due to their design and purpose, STXs apply mainly to platoon- and company-size units. This does not preclude the development of STXs or similar exercises for unit staffs and headquarters. They combine "how-to-train" specifics for selected collective tasks with the "what-to-train" specifics found in doctrinal manuals and other documents.

G-7. Performance-oriented training.

a. Performance-oriented training emphasizes hands-on performance. Training is conducted to ensure a level of proficiency, rather than established number of hours of required training.

b. Performance-oriented training has precise training objectives and makes good use of resources. Training objectives serve as the basis for preparing, conducting, and evaluating all training.

G-8. Evaluation.

a. A training program must include an evaluation of training (planned and conducted). An evaluation plan must:

(1) Provide immediate feedback to the trainer.

(2) Recognize effective trainers.

(3) Evaluate all units in the conduct of training to provide for a comparison of training programs.

(4) Be flexible to unit needs; provide general and specific evaluations based on the level of training within the unit.

(5) Be in effect at all times.

b. Major considerations for training evaluations include:

(1) Ranger unit commanders establish and maintain an effective internal and external evaluation program.

(2) The director of each staff element is responsible for training evaluations within his area of staff responsibility with emphasis on low-density MOS groups.

(3) The operations officer of the ranger organization develops and validates training and evaluation plans.

c. Evaluation of training must be done vertically from the major task (mission) down through subtasks (collective tasks) to the leader and individual tasks. In selection of individual tasks for evaluation, priority should be given to critical individual tasks that impact on the success of the collective task or mission.

d. The critique and after-action review, after each major training event, are the most important aspects of training. The after-action review takes place immediately after completion of a training task (individual or collective). The individual(s) self-analyzes and critiques the training event. Then after a formal critique of the training, the individuals) trains to standard (not to time).

G-9. Field training.

a. **Considerations.** Ranger commanders schedule training under field conditions to permit maximum training opportunities. Field training emphasis is on mastery of a series of individual (soldier and leader) tasks and collective training tasks. These tasks are repeated until established standards are met. The goal of ranger field training is to ensure realistic, demanding, and meaningful training.

(1) When training is conducted under conditions of moderate stress, most combat critical skills are learned quicker and retained longer. (See Figure G-3 on the following page.) Stress can be incorporated in a training situation by physical deprivation (sleep, food), by live-fire exercises, by operation against an OPFOR under simulated battlefield conditions for an extended period, by the intensity of activity including demands for decision making, or by physically demanding activities. Ranger units train under stress so that they will react quickly in combat.

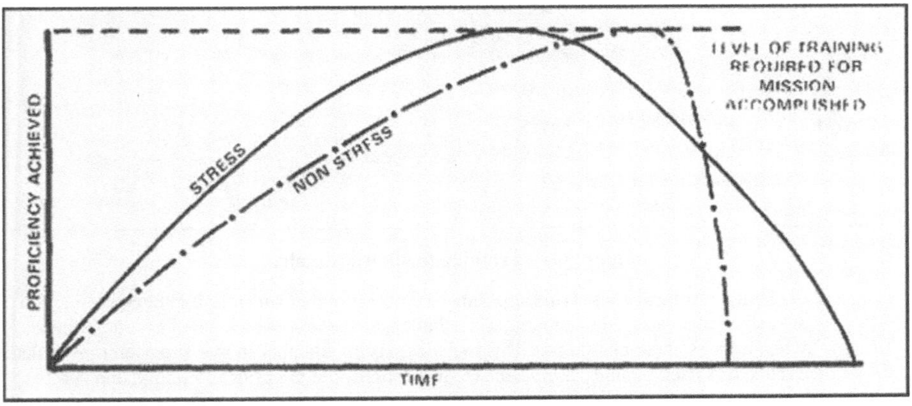

Figure G-3. The Army training and evaluation program.

(2) Antagonistic training situations are two-sided exercises conducted by opposing elements. Each element has a training objective (task and standards). The opposing force has a corresponding countertask. This training may be progressive--for example, starting with a

crawl-through, then a walk-through, and progressing to both elements operating at a fast pace. The use of multiple integrated laser equipment system (MILES) greatly enhances task and countertask training. The units should also be considered for a live-fire exercise against a realistic OPFOR target array when the basics have been mastered.

b. **Methods of training.**

(1) **Considerations.** The chance for success in combat can be increased by becoming proficient in training situations that depict realistic combat conditions. (See Figure G-4.) Meeting the standards of a realistic live-fire training exercise assures the commander that his unit could perform well in combat. Figure G-4 shows the relationship between realistic training and the challenge to the trainer.

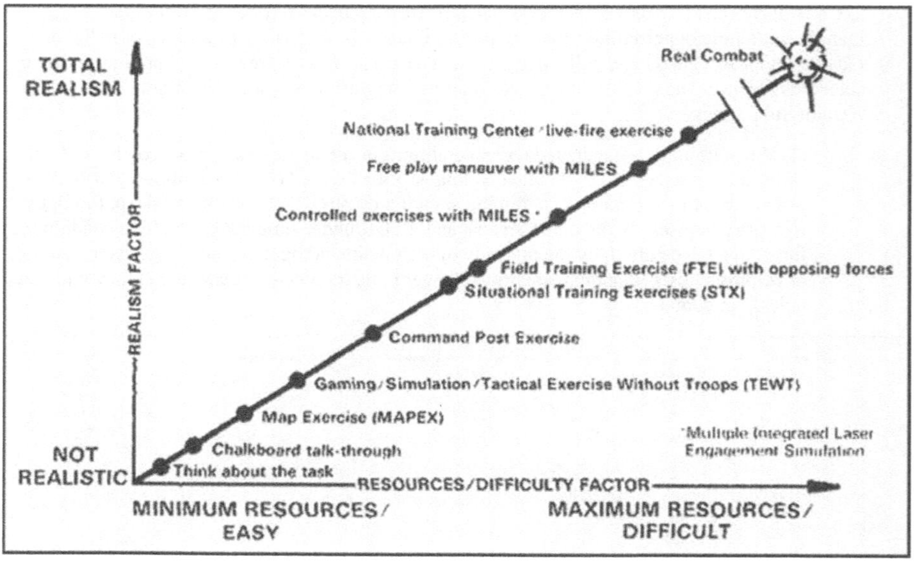

Figure G-4. Training proficiency scale.

(2) **Ranger cohesiveness training.** Ranger units promote teamwork by undergoing tough, stressful, often hazardous training. This challenges the unit, which gains in cohesiveness when it meets these challenges. This includes serving together in live-fire exercises, athletic events, precision parades, high-standard inspections, tough physical training, and combat.

(3) **Field training exercise.** An FTX is training normally conducted under simulated, realistic combat conditions. This training normally exercises command and control systems and the administrative and logistical needs of sustained combat. Stress training is part of this exercise, focusing on live-fire and MILES drills and STXS.

(4) **Combined arms training.** Combined arms training should include all means of support such as tactical air, armed helicopters, and indirect fire with emphasis on fire support coordination and airspace control.

(5) **Live-fire exercises.** Including live-fire exercises in drills and STXs offers a chance to evaluate units in their response to simulated combat missions. Ranger units train under pressure and always anticipate being selected to conduct a live-fire exercise. Live-fire exercises develop confidence and esprit, and reinforce unit discipline.

(6) **Opposing force.** Committing one unit against another adds a competitive spirit to a field training exercise (FTX). The use of MILES and squad combat operations exercises (simulation) (SCOPES) provides realism to two-sided field exercises. Trainers should use OPFOR equipment and training devices available, as well as making their own devices. (See FM 30-103, Appendix R, for more information.)

c. **Key functional training areas.**

(1) **Land navigation and patrolling.** Day and night navigation and patrolling exercises must be emphasized. Orienteering improves these skills, as well as physical fitness.

(2) **Movement techniques.** Infantry combat has two major components--to move and to hit the enemy first. Ranger units must be skilled in the use of traveling, traveling overmatch, and bounding overmatch. Units in contact must be skilled in moving under fire.

(3) **Fighting** positions. Training includes the building of concealed fighting positions.

(4) **Mine and countermine training.** Training includes how to defend against hostile mines, booby traps, and other demolitions. The identification, installation, neutralization, and removal of US mines and devices is also stressed.

(5) **Limited visibility operations.** Ranger units must be skilled in limited visibility operations and the use of night vision and thermal devices.

(6) **Operation security.** Operation security includes deception, physical security, signal security (SIGSEC), and information security. It is taught in all training and at all levels. Although all areas of OPSEC must be measured in training, techniques of camouflage and SIGSEC need special emphasis. (See AR 530-1 and TC 100-1 for more information.)

G-10. Professional development.

Ranger units at battalion and regimental levels conduct many officer and NCO professional development programs. This ensures that all officers and NCOs are skilled in their field, confident, and wellrounded. Professional readings and book reports, tactics seminars, and tactical exercises without troops are good development methods.

G-11. Physical training.

a. This program ensures that each ranger maintains the physical ability and stamina needed to do his mission. Other objectives are to promote unit cohesiveness and discipline. These are met by combining individual testing, formation, running, athletic competition, weight control, proper diet, teamwork drills, and freedom from drug and alcohol abuse. Mental fitness and toughness come from experience and success. Good physical conditioning increases a unit's chances of success.

b. Physical training (PT) programs are designed to include the elements of overload, progression, balance, variety, and regularity.

c. Ranger unit physical fitness standards are:

(1) Score 80 points for each Army physical readiness test (APRT) event and do six chinups.

(2) Pass the ranger swimming test.

(3) Complete an 8-kilometer run in 40 minutes.

(4) Complete a 12-kilometer road march in three hours (with rucksack, helmet, and weapon).

(5) Meet the Army height and weight standards.

G-12. Environmental and specialized training.

a. Rangers conduct training in environments as close to real combat missions as possible. This conditions rangers to survive and fight in a specific environment. Such training includes combat-critical tasks and technique training for that environment.

b. Each ranger battalion's environmental and specialized training includes at least the following:

(1) Jungle (once every 12 months).

(2) Desert (once every 12 months).

(3) Urban (once every 6 months).

(4) Extreme cold (once every 18 months).

(5) Mountain (once every 12 months).

(6) Amphibious (once every 18 months).

G-13. Ranger indoctrination program.

a. The RIP is a training and selection program. It is run by specially selected NCOs from the regimental headquarters. It has as its mission the training and indoctrination of newly assigned soldiers. It physically toughens and mentally sharpens the newly assigned soldier. It teaches basic skills and techniques. The RIP identifies and eliminates in three weeks any applicant who does not show dedication, motivation, physical fitness, and emotional stability.

b. An indoctrination program for newly assigned ranger-qualified officers and NCOs prepares them to assume duty as leaders within the ranger regiment. These personnel monitor and participate in RIP training, as well as undergoing individual ranger leadership training. The RIP provides them an overview of the regiment's history, its mission, and its goals. (See Appendixes C,

D, and E for further information about the history and goals of the ranger regiment.) The RIP also provides a review of the skills required in a ranger unit. Refresher training in airborne and pathfinder operations, use of indirect fires, and light infantry doctrine is conducted. During RIP all newly assigned officers and NCOs must meet the ranger physical fitness standard.

c. The RIP cadre also conducts a preranger program to prepare soldiers to attend the US Army Ranger School. It is a physically demanding course that teaches troop-leading procedures, operation orders, and small unit patrol actions.

Glossary

The following acronyms, abbreviations, and definitions are provided to aid the reader in understanding this publication.

Section I

ACRONYMS AND ABBREVIATIONS

AAT	Army assault team
ABCCC	airborne battlefield command and control center
ACL	allowable cargo load
ADA	air defense artillery
AFFOR	the USAF element of a joint task force
AGL	above ground level
AKZ	armor kill zone
ALCC	airlift control center
ALCE	airlift control element
AM	amplitude modulation
AMC	air mission commander
APERS	antipersonnel
APRT	Army physical readiness test
ARFOR	the Army element of a joint task force
ARSOC	Army special operations command
ARSOF	Army Special Operations Forces
ASG	area support group
ASPC	all-source production center

ATGM	antitank guided missile
AWADS	adverse weather aerial delivery system
BCE	battle coordination element
BTMS	battalion training management system
CA	civil affairs
CAS	close air support
CCT	combat control team
CDS	container delivery system
CE	communications-electronics
CI	counterintelligence
CINC	commander in chief
COMINT	communications intelligence
COMSEC	communications security
CONUS	continental United States
COSCOM	corps support command
CP	command post
CS	combat support; a chemical agent
CSS	combat service support
DACG	departure airfield control group
DACO	departure airfield control officer
DASC	direct air support center
DMDG	digital message device group
DOD	Department of Defense
DS	direct support
DSU	direct support unit
DZ	drop zone
EAC	echelons above corps

EACIC	Echelons Above Corps Intelligence Center
ECCM	electronic counter-countermeasure
ECM	electronic countermeasures
E&E	escape and evasion
EEFI	essential elements of friendly information
EOC	emergency operations center
EW	electronic warfare
FARP	forward area rearm and refuel point
FASCAM	family of scatterable mines
FAX	facsimile transmission
FDC	fire direction center
FEBA	forward edge of the battle area
FFL	friendly front lines
FIST	fire support team
FLOT	forward line of own troops
FM	field manual; frequency modulation
FO	forward observer
FORSCOM	US Army Forces Command
FPF	final protective fire
FPL	final protective line
FRAGO	fragmentary order
FSA	fire support area
FSCOORD	fire support coordination or coordinator
FSE	fire support element
FSO	fire support officer
FSS	fire support station
FTX	field training exercise

G3	Assistant Chief of Staff, Operations
GS	general support
GSR	ground surveillance radar
HAARS	high-altitude, airdrop resupply system
HAHO	high-altitude, high-opening
HALO	high-altitude, low-opening
HE	high explosive
HF	high frequency
HHC	headquarters and headquarters company
HSLLADS	high-speed, low-level airdrop system
HUMINT	human intelligence
IAW	in accordance with
IFF	identification-friend-or-foe
IP	initial point
IRP	initial rally point
ISB	intermediate staging base
JACC/CP	joint airborne communications center/command post
JCS	Joint Chiefs of Staff
JCSE	joint communications support element
JFC	joint force commander
JSOC	joint special operations command
JTB	joint transportation board
JTF	joint task force
kw	kilowatt
LAPES	low altitude parachute extraction system
LAW	light antitank weapon
LNO	liaison officer

LOC	logistical operations center
LRRP	long-range reconnaissance patrol
LRSU	long-range surveillance unit
LTD	laser target designator
LZ	landing zone
MAC	Military Airlift Command
MANPADS	man-portable air defense system
MCC	movement control center
METT-T	mission, enemy, terrain, troops and time available
MI	military intelligence
MILES	multiple integrated laser engagement system
mm	millimeter
MMC	Materiel Management Center
MMEE	minimum mission-essential equipment
MOS	military occupational specialty
MOUT	military operations on urbanized terrain
MRLS	multiple rocket launch system
MTP	mission training plan
NBC	nuclear, biological, chemical
NCA	National Command Authority
NCO	noncommissioned officer
NOD	night observation device
OCOKA	observation, concealment, obstacles, key terrain, and avenues of approach
OCONUS	outside continental United States
OPCOM	operational command
OPCON	operational control
OPFOR	opposing force

OPORD	operation order
OPSEC	operations security
ORP	objective rally point
PAC	personnel administration center
PDF	primary direction of fire
PIR	priority intelligence requirements
PLL	prescribed load list
POM	preparation for oversea movement
PSYOP	psychological operations
PT	physical training
PZ	pickup zone
QRE	quick-reaction element
RATELO	radiotelephone operator
REMAB	remote marshalling base
RFL	restrictive fire line
RIP	ranger indoctrination program
RP	rally point; release point
RPG	rocket-propelled grenade
RRF	ranger ready force
RRP	reentry rally point
RSE	ranger support element
RSTA	reconnaissance, surveillance, and target acquisition
S1	adjutant
S2	intelligence officer
S3	operations and training officer
S4	supply officer
S5	civil affairs officer

SAR	search and rescue
SATCOM	satellite communications)
SCIF	sensitive compartmented information facility
SCOPES	squad combat operations' exercises (simulation)
SCUBA	self-contained underwater breathing apparatus
SEAD	suppression of enemy air defense
SEMA	special electronic mission aircraft
SERE	survival, evasion, resistance, and escape
SF	special forces
SHORAD	short-range air defense
SIGSEC	signal security
SKE	station-keeping equipment
SLAR	side-looking airborne radar
SOA	special operations aviation
SOCB	special operations communications battalion
SOCOM	special operations command (usually, 1st SOCOM)
SOP	standing operating procedures
SOSB	special operations support battalion
SOTF	special operations task force
SPL	support platoon leader
STX	situational training exercise
TA	theater Army
TAACOM	Theater Army Area Command
TAC	Tactical Air Command
TACP	tactical air control party
TACSAT	tactical satellite terminal
TAMMC	Theater Army Materiel Management Center

TASG	theater Army support group
TDA	table of distribution and allowances
TOA	transfer of authority
TOC	tactical operations center
TOE	table(s) of organization and equipment
TRP	target reference point
TTY	teletype
UHF	ultrahigh frequency
US	United States
USAF	United States Air Force
USMC	United States Marine Corps
USN	United States Navy
VHF	very high frequency
WP	white phosphorus
XO	executive officer

Section II

DEFINITIONS

ADVANCE FORCE. An operation requiring the immediate deployment of US forces with the intent to negate a threat before the deployment of follow-on forces is required.

AIRLAND BATTLE DOCTRINE. An approach to military operations that realizes the full potential of US forces. Two notions (extending the battlefield and the ability to integrate conventional, nuclear, chemical, and electronics means) are combined to describe a battlefield where the enemy is attacked to the full depth of his formation.

AREA OF INFLUENCE. A geographical area wherein a commander is directly capable of influencing operations by maneuver or fire support systems normally under his command or control.

AREA OF INTEREST. That area of concern to the commander, including the area of influence, areas adjacent thereto, and extending into enemy territory to the objectives of current or planned operations. This area also includes areas occupied by enemy forces who would jeopardize the accomplishment of the mission.

AREA OF OPERATIONS. That portion of an area of conflict necessary for military operations. Areas of operations are geographical areas assigned to commanders for which they have responsibility and in which they have authority to conduct military operations.

CHALK. One specific aircraft load.

CLANDESTINE OPERATIONS. Activities to accomplish intelligence, counterintelligence, and other similar activities sponsored or conducted by governmental departments or agencies, in such a way as to ensure secrecy or concealment. (They differ from covert operations in that emphasis is placed on concealment of the operation rather than on concealment of the identity of the sponsor.)

COMMUNICATIONS INTELLIGENCE (COMINT). The interception and processing of foreign communications passed by radio, wire, or other electromagnetic means, and by the processing of foreign encrypted communications, however transmitted. Interception comprises search, intercept, operator identification, signal analysis, traffic analysis, crypto-analysis, study of plain text, the fusion of these processes, and reporting the results.

COMMUNICATIONS SECURITY. The protection resulting from all measures designed to deny to unauthorized persons information of value that might be derived from the possession and study of telecommunications, or to mislead unauthorized persons in their interpretations of the results of such study. COMSEC includes crypto-security, physical security, and transmission security.

COUNTERINTELLIGENCE. That aspect of intelligence activities, both offensive and defensive, designed to detect and neutralize or destroy the effectiveness of hostile foreign intelligence activities and to protect information against espionage, personnel against subversion, and installations and materiel against sabotage.

COVERT OPERATIONS. Operations that are so planned and executed as to conceal the identity of or permit plausible denial by the sponsor. They differ from clandestine operations in that emphasis is placed on concealment of identity of the sponsor rather than on concealment of the operation.

DECEPTION. Those measures designed to mislead the enemy by manipulation, distortion, or falsification of evidence to induce him to react in a manner prejudicial to his interests.

DEMONSTRATION. A show of force in an area where a decision is not sought. It is similar to a feint but does not make contact with the enemy.

DRILL. A drill focuses on a "chunk" or "slice" of battle and is targeted at small units. Drills are critical collective tasks that require a high degree of proficiency from a small unit. Generally, small units "drill" collective tasks that require rapid responses by unit members in the absence of detailed orders from unit leaders.

ELECTRONIC COUNTERMEASURES. That division of electronic warfare involving actions taken to prevent or reduce an enemy's effective use of the electromagnetic spectrum.

ELECTRONIC INTELLIGENCE (ELINT). The technical and intelligence information derived from foreign noncommunications electromagnetic radiations emanating from other than atomic detonations or radioactive sources.

ELECTRONIC WARFARE. Military action involving the use of electromagnetic energy to determine, exploit, reduce, or prevent hostile use of the electromagnetic spectrum and action that retains friendly use of the electromagnetic spectrum.

ELECTRONIC WARFARE SUPPORT MEASURES (ESM). That division of EW involving actions taken to search for, intercept, identify, and locate sources of radiated electromagnetic energy for the purpose of immediate threat recognition. ESM provides a source of information needed for immediate action involving electronic countermeasures, electronic counter-countermeasures, avoidance, targeting, and other tactical employment of forces.

ESPIONAGE. Actions directed toward the acquisition of information through clandestine operations.

EVACUATION/SAFEGUARD. The use of US forces to evacuate US citizens or allies from a hostile or potentially hostile area, or to safeguard US personnel or property.

EXFILTRATION. The removal of personnel or units from areas under enemy control by stealth, deception, surprise, or clandestine means. Exfiltration is a method of extracting ranger units.

EXTRACTION. The removal of personnel or units from areas under enemy control by any means, to include exfiltration, airland, airmobile, or amphibious.

FEINT. A diversionary supporting attack conducted to draw the enemy's attention from the main effort. It is normally executed by brigades and small units. Feints are usually shallow, limited objective attacks that go in before or during the main attack.

GUERRILLA WARFARE. Military and paramilitary operations conducted in enemy-held or hostile territory by irregular and predominantly indigenous forces.

HIGH-INTENSITY CONFLICT. A war between the US and any other nation or nations in which both belligerents use the most modern technology and all resources in intelligence, mobility, firepower (including NBC), command, communications and control, and service support.

HUMAN INTELLIGENCE. Intelligence derived from information provided by human resources.

INFILTRATION. The movement through or into an area of territory occupied by either friendly or enemy troops or organizations. The movement is made either by small groups or by individuals, at extended or irregular intervals. When used in conjunction with the enemy, it implies that contact is avoided. Infiltration is a subset of insertion.

INSERTION. Placement of troops and equipment into an operational area by any means.

INTERDICT. To isolate or seal off an area by military means; to prevent, hinder, or delay the use of an area or route by enemy forces.

INTERDICT LINE OF COMMUNICATIONS. An attack to seal off an area or to deny use of a route or approach.

JOINT TASK FORCE. A JTF may be constituted and designated by the Secretary of Defense or by the commander of a unified command, specified command, or an existing JTF. Normally, it performs missions having specific, limited objectives or missions of short duration. It dissolves when it has achieved its purpose. The joint task force commander is responsible to the commander, known as the establishing authority, who created the JTF. The JTF is composed of elements of two or more services operating under a single JTF commander. The JTF commander has operational control over the entire force and may have direct command of his own staff with representatives from the other services. He exercises logistics coordination or control only as necessary to meet his subordinate commanders' logistics needs.

LIAISON. That personal contact or communication maintained between elements of military forces to ensure mutual understanding and unity of purpose and effort.

LINE OF COMMUNICATIONS. All routes (land, water, and air) that connect an operating military force with one or more bases of operations and along which supplies and military forces move.

LOW-INTENSITY CONFLICT. Limited politico-military operations conducted to achieve political, social, economic, or psychological objectives. It is generally confined to a specific geographic area and is often characterized by operational constraints imposed on ranger units by political and economic considerations. It may involve intermittent combat against regular threat military forces, but it more often involves conflict with irregular, security, or paramilitary forces of a country or group involved in an operation that is counter to the best interests and national security of the United States. It is characterized by relatively low levels of combat and constraints placed on the weaponry, tactics, and movements of the ranger force. It may involve counterterrorist or antiterrorist operations and generally occurs before a pro forma declaration of war.

MID-INTENSITY CONFLICT. War, declared or undeclared, between the US and the regular forces of one or more nations and their respective allies, if any, in which the belligerents use the most modern technology and all resources in intelligence, mobility, firepower (excluding nuclear, chemical, and biological weapons), command, communications and control, and support. They are used for limited objectives under definitive policy limitations as to the extent of destructive power that can be used or the extent of geographical area that might be involved.

MILITARY STRATEGY. Art and science of using the armed forces of a nation to secure the objectives of national policy by application of force or the threat of force. Military strategy sets the fundamental conditions for operations.

NATIONAL COMMAND AUTHORITY. This includes the President and Secretary of Defense or the duly deputized alternates or successors. Commonly referred to as NCA.

OPERATIONAL COMMAND/CONTROL. Operational command (OPCOM) and operational control in joint force terminology both refer to the authority exercised by joint commanders over subordinate service components. Those terms are not interchangeable in joint operations. OPCOM applies to the authority exercised by commanders of unified commands. OPCON is the authority that subordinate joint task force commanders exercise in the conduct of specific operations. The authority that military departments exercise over their respective components is commonly referred to as command less operational command.

OPERATIONAL LEVEL OF WAR. The operational level of war uses available military resources to attain strategic goals within a theater of war. It is the theory of larger unit operations. It also involves planning and conducting campaigns. Campaigns are sustained operations designed to defeat an enemy force in a specified space and time with simultaneous and sequential battles. The disposition of forces, selection of objectives, and actions taken to weaken or to outmaneuver the enemy all set the terms of the next battle and exploit tactical gains. They are all part of the operational level of war. In AirLand Battle doctrine, this level includes the marshalling of forces and logistical support, providing direction to ground and air maneuver, applying conventional and nuclear fires in depth, and employing unconventional and psychological warfare.

OPERATIONS SECURITY. Those measures designed to protect information concerning planned, ongoing, and completed operations from unauthorized disclosure. It includes all actions a command takes to deny the enemy information about friendly units and their operations.

ORDER OF BATTLE (OB). The identification, strength, command structure, and disposition of the personnel, units, and equipment of any military force.

RAID. Usually a small-scale operation, involving a swift penetration of hostile territory to secure information, confuse the enemy, or destroy his installations. It ends with a planned withdrawal upon completion of the assigned mission.

RANGER MISSION. To conduct strike, raid, or special light infantry operations. Execution of ranger missions is normally directed by the National Command Authority. Rangers also support operational-level commanders during operations of limited duration when allocated or assigned by the theater commander.

RECONNAISSANCE. A mission undertaken to obtain, by visual observation or other detection methods, information about the activities and resources of an enemy or potential enemy; or to secure data concerning the meteorological, hydrographical, or geographical characteristics of a particular area.

SEARCH AND RESCUE. The use of aircraft, surface craft, submarines, specialized rescue teams, and equipment to search for and rescue personnel in distress on land or at sea.

SEIZURE. As an operational purpose, the capture of a voluntarily restricted portion of an enemy-controlled territory. Capture of an isolated land mass such as an island is usually categorized as a seizure.

SEIZE KEY FACILITY. An attack on an installation with the intent of holding it for a short period. Normally, rangers would make the assault to seize and secure an installation for follow-on forces.

SERVICE COMPONENT. Each service component commander is responsible for recommending the proper use of his forces and for accomplishing operational tasks assigned by the joint commander. He is also responsible for his service in matters of internal administration and discipline; training in own service doctrine, techniques, and tactics; designation of specific units to meet joint requirements; logistics functions normal to the component; tactical employment of service component forces; and service intelligence matters.

SHOW OF FORCE. A mission carried out to demonstrate US resolve, whereby US forces are deployed to defuse a situation that may be harmful to US interests or national objectives.

SIGNAL INTELLIGENCE (SIGINT). A generic term that includes COMINT and ELINT. Activities concerned with the collection and processing of foreign electromagnetic emissions for the purpose of producing intelligence information.

SITUATIONAL TRAINING EXERCISE. An STX is a short, mission-oriented exercise in which a group of closely-related collective tasks are trained, using a doctrinally preferred method of executing these tasks to established standards. An STX is drill-like in nature, but it is more complex and flexible than a drill. It usually involves a larger unit than a drill.

SPECIAL OPERATIONS. Military operations conducted by specially trained, equipped, and organized DOD forces against strategic or tactical targets in pursuit of national, military, political, economic, or psychological objectives. Special operations may support conventional military operations, or they may occur independently when the use of conventional forces is either inappropriate or infeasible. Sensitive peacetime operations, except for training, are normally authorized by the NCA and conducted under the direction of the NCA or designated commander. Special operations may include unconventional warfare, counter-terrorist operations, collective security, psychological operations, and civil affairs measures.

SPECIAL OPERATIONS FORCES. All civil affairs, psychological operations, ranger (active component), and special forces units within the US Army, US Army Reserve, and National Guard.

SPECIAL OPERATIONS COMMAND. The major subordinate command to FORSCOM exercising command and control of all active component Army special operations forces not placed under the operational command of unified or specified commands.

STRATEGIC LEVEL OF WAR. Military strategy uses the armed forces of a nation to secure the objectives of national policy by applying force or the threat of force. Military strategy sets the fundamental conditions for operations.

STRIKE. An attack that is intended to inflict damage on, seize, or destroy an objective. Sometimes referred to as direct action missions.

SURVEILLANCE. The systematic observation of aerospace, surface or subsurface areas, places, persons, or things by visual, aural, electronic, photographic, or other means.

TACTICAL. Pertaining to the employment of units in combat.

TACTICAL LEVEL OF WAR. Tactics are the specific techniques smaller units use to win battles and engagements that support operational objectives. Tactics employ all available combat, combat support, and combat service support. Tactics involve the movement and positioning of forces on the battlefield in relation to the enemy, the provisioning of fire support, and the logistical support of force before, during, and following engagements with the enemy. At corps and division, operational and tactical levels are not clearly separable, but they are guided by the same principles. An operation designed to defeat any enemy force in an extended area does so through operational maneuver and a series of tactical actions.

TACTICAL COMBAT INTELLIGENCE. That knowledge of the enemy, weather, and geographical features needed by a commander in the planning and conduct of combat operations. It is derived from the interpretation of information on the enemy and the environment.

TARGET ANALYSIS. A detailed and systematic examination of processed intelligence to identify and locate targets.

UNIFIED COMMAND. A unified command is established and designated by direction of the President. The command is organized under a single unified CINC and has a broad, continuing mission. The CINC is responsible to the Secretary of Defense for accomplishing his broad and continuing mission and operates under the strategic and operational direction of the JCS. A unified command is composed of assigned forces of two or more services. These forces, as service components, are under the operational command of the unified commander. The CINC exercises operational command authority through service component commanders. When tasked by higher authority, he may establish a subordinate JTF to conduct specific missions. Under emergency conditions, the CINC has the authority to use all facilities and supplies of assigned forces to accomplish his mission.

References

REQUIRED PUBLICATIONS

Required publications are sources that users must read in order to understand or to comply with this publication.

Field Manuals (FM)

FM 7-8	The Infantry Platoon and Squad (Infantry, Airborne, Air Assault, Ranger)
FM 7-10	The Infantry Rifle Company (Infantry, Airborne, Air Assault, Ranger)
FM 7-20	The Infantry Battalion (Infantry, Airborne, and Air Assault)

RELATED PUBLICATIONS

Related publications are sources of additional information. They are not required in order to understand this publication.

Army Regulations (AR)

AR 350-1	Army Training

Field Manuals (FM)

FM 10-52-1	Commander's Handbook for Water Usage in Desert Operations
FM 10-501	Airdrop of Supplies and Equipment: Rigging Containers
FM 10-512	Airdrop of Supplies and Equipment: Rigging Typical Supply Loads
FM 10-542	Airdrop of Supplies and Equipment: Rigging Loads for Special Operations
FM 10-547	Airdrop of Supplies and Equipment: Rigging the High Speed Aerial Delivery Container CTU-2/A
FM 10-553	Airdrop of Supplies and Equipment: Rigging Ammunition
FM 12-3-4	Echelons Above Corps (EAC) Personnel and Administrative Doctrine

FM 14-6 Comptroller/Finance Services in Theaters of Operations

FM 19-15 Civil Disturbances

FM 20-11-1 Military Diving (Volume 1)

FM 20-11-2 Military Diving (Volume 2)

FM 21-20 Physical Training Program

FM 21-26 Map Reading

FM 21-75 Combat Skills of the Soldier

FM 21-76 Survival, Evasion, and Escape

FM 21-77A(S) Joint Worldwide Evasion and Escape Manual

FM 22-9 Soldier Performance in Continuous Operations

FM 23-67 Machinegun, 7.62-mm, M60

FM 25-1 Training

FM 25-2 Unit Training Management

FM 25-3 Training in Units

FM 25-4 How to Conduct Training Exercises

FM 25-5 Training for Mobilization and War

FM 25-7 Training Ranges

FM 26-2 Management of Stress in Army Operations

FM 29-47 Main Supply and Service Company, Forward Supply and Service Detachments, Supply and Service Battalion, Airborne Division

FM 29-51 Division Supply and Field Service Operations

FM 30-5 Combat Intelligence

FM 30-17 Counterintelligence Operations

FM 31-18 Long-Range Reconnaissance Ranger Company

FM 31-19 Special Forces Military Free-Fall Parachuting

FM 1-24 Special Forces Air Operations

FM 31-25	Special Forces Waterborne Operations
FM 31-70	Basic Cold Weather Manual
FM 34-1	Intelligence and Electronic Warfare Operations
FM 34-60A	Counterintelligence Operations (U)
FM 55-9	Unit Air Movement Planning
FM 57-38	Pathfinder Operations
FM 57-220	Basic Parachuting Techniques and Training
FM 63-3J	Combat Service Support Operations--Corps
FM 63-4	Combat Service Support Operations--Theater Army Area Command
FM 63-5	Combat Service Support Operations--Theater Army
FM 90-3	Desert Operations
FM 90-4	Airmobile Operations
FM 90-5	Jungle Operations
FM 90-6	Mountain Operations
FM 90-8	Counterguerilla Operations
FM 90-10	Military Operations on Urbanized Terrain (MOUT)
FM 90-10-1	An Infantryman's Guide to Urban Combat
FM 100-5	Operations
FM 100-20	Low Intensity Conflict
FM 100-26	The Air-Ground Operations System
FM 101-5	Staff Organization and Operations
FM 101-10-1	Staff Officers' Field Manual: Organization, Technical and Logistical Data (U)

Firing Table (FT)

FT 60-P-1	Firing Tables for Mortar, 60-mm: M224, Firing Cartridge, HE, M720, Cartridge, HE, M49A4, Cartridge, TP, M5OA3, Cartridge, WP, M302AI, Cartridge, Illuminating, M83A3

Soldier's Training Publication (STP)

STP 7-11B24-SM	Soldier's Manual MOS 11B, Infantryman, Skill Level 2/3/4

Technical Manuals (TM)

TM 10-1670-267-12&P	Operator's and Organizational Maintenance Manual Including Repair Parts and Special Tools List for High Altitude Airdrop Resupply System: 500 Pound Capacity

Training Circulars (TC)

TC 23-14	Sniper Training and Employment
TC 23-90	Mortar Training
TC 24-1	Communications-Electronics Operation Instructions, the CEOI
TC 31-20-5(C)	Special Forces Communications (U)
TC 90-6-1	Military Mountaineering

Other Related Publications

King, Michael J. Dr., *Leavenworth Papers, No. 11*, "Rangers Selected Combat Operations in World War II," Combat Studies Institute, USACGSC, Fort Leavenworth, KS

PROJECTED PUBLICATIONS

Projected publications are sources of additional information that are scheduled for printing but are not yet available. Upon print, they will be distributed automatically via pinpoint distribution. They may not be obtained from the USA AG Publications Center until indexed in DA Pamphlet 310-1.

Field Manuals (FM)

FM 7-70	Light Infantry Squad/Platoon
FM 7-71	Light Infantry Company
FM 7-72	Light Infantry Battalion
FM 7-73	Light Infantry Brigade

FM 23-36 60-mm Mortar, M224